Lockheed C-5 Galaxy

Lockheed C-5

GALAXY

Chris Reed

Schiffer Military History
Atglen, PA

Dedication

To All Airlifter Crews

Photo Credits

Lockheed Martin Aeronautical Systems
Andreas Spagna
Nick Challoner (www.netmontage.co.uk/airspeed/)
Paul Hart
Paul Osborne/UK Airshow Review (www.airshows.co.uk)
Terry Lee/Aviation Consultants, Ltd, UK (www.aviation-consultants.com)
Tim Doherty
Robert M. Robinson
Harry Heist/Air Mobility Command Museum (www.amcmuseum.org)
Arno J.A.H. Cornelissen
Masanori Ogawa
Hans Heijdenijk
Andy Thomson
Simon N.J. Edwards

Other Assistance
Bill Gardner
Lockheed Martin Aeronautical Systems/Susan A. Miles
Howard J. Curtis
Robert Acosta
Jim Benson
Francesco Missarino (members.tripod.com/MIX1976)

Book Design by Ian Robertson.

Library of Congress Catalog Number: 00-108227

Printed in China.
ISBN: 0-7643-1205-7

We are interested in hearing from authors with book ideas on related topics.

Published by Schiffer Publishing Ltd.
4880 Lower Valley Road
Atglen, PA 19310
Phone: (610) 593-1777
FAX: (610) 593-2002
E-mail: Schifferbk@aol.com.
Visit our web site at: www.schifferbooks.com
Please write for a free catalog.
This book may be purchased from the publisher.
Please include $3.95 postage.
Try your bookstore first.

In Europe, Schiffer books are distributed by:
Bushwood Books
6 Marksbury Avenue
Kew Gardens
Surrey TW9 4JF
England
Phone: 44 (0) 20 8392-8585
FAX: 44 (0) 20 8392-9876
E-mail: Bushwd@aol.com.
Free postage in the UK. Europe: air mail at cost.
Try your bookstore first.

Contents

Chapter One

Setting the Stage

For many years the world's largest military aircraft, and still the largest type in the USAF inventory more than thirty years after it first appeared, Lockheed's C-5 Galaxy had a troubled development and early history that came close to destroying the company that created it, yet went on to become a cornerstone of the USAF's ability to project power worldwide.

Although large transport aircraft designs date back to the First World War, World War II really defined the need for long-range strategic airlifters. While the majority of military logistics still flowed by sea, the ability of aircraft to quickly bring significant loads of men, weapons, and materials into theaters, and even directly into battle zones spurred the development of modern military airlifters. However, for the most part, the USAAF's wartime air transport capability centered around the C-46, C-47, and C-54 versions of the Curtiss CW-20, as well as the Douglas DC-3 and DC-4 airliners. These workhorses were produced by the thousands and saw heavy use all over the globe, but all three were at their core commercial aircraft, ill-suited for the loading and carriage of outsized and indivisible loads, such as large military vehicles.

For example, the C-54 Skymaster's maximum 36,000 lb payload limit was greater than the weight of an M5 Stuart light tank, but the Skymaster's fuselage could not possibly accomodate a Stuart; even if it would, the portside loading door on the aft fuselage was not large enough to pass such a load.

Planning for a heavy strategic airlifter actually began not long after Pearl Harbor, although wartime demand for aircraft already in production meant that the type would not see active service in the conflict. Douglas, which had its hands full with DC-3 and DC-4 production, nevertheless was able to design the first of the American "monster airlifters," the C-74 Globemaster. In many ways resembling the smaller and slightly later DC-6/C-118 Liftmaster, the Globemaster was distinctly a product of the company's Santa Monica division, with a circular cross-section fuselage, low-mounted wing, and a typically "Douglas" tail. Perhaps the type's most no-

One of the largest of the wartime transport designs, the Lockheed Constellation was, like many types before it, designed as a passenger transport, and was not ideally suited for carrying many military-type loads, such as large vehicles and construction equipment. (Chris Reed)

The Douglas C-133 Cargomaster provided the USAF with much of its outsized lift capability up until the type was retired in 1971. This particular example, a C-133A, set time to climb records in 1958, and is on display at the USAF Museum. (Chris Reed)

The very first Starlifter, 61-2775. Unveiled by President Kennedy, -2775 spent its more than thirty year career as a test machine, having the "temporary test" designation JC-141A before becoming a "permanent test" NC-141A. (Air Mobility Command Museum Photo)

ticeable external feature, apart from its size, were the separate "bug eye" canopies for the pilot and copilot, a configuration that was also a feature of the never-built XB-31 bomber.

Powered by four of the new large Pratt & Whitney R4360-27 Wasp Major 28-cylinder radial engines, the C-74 could carry 48,000 lbs of cargo (four times the capacity of the C-54) or 125 troops. Like so many other military aircraft programs, the C-74 suffered severe cutbacks at the end of the war, with all but fourteen aircraft out of the original order of fifty being canceled. The few Globemasters that did see service made some notable flights; one was the first to carry more than 100 passengers across the Atlantic, taking 103 passengers and crew to the base at Marham, UK, non-stop from Mobile, Alabama. Another took part in the Berlin Airlift, bringing in construction equipment to aid in the building of airflieds in the Western Zones.

The C-74 was retired by the USAF in 1956, and although three or four examples had second careers as commercial haulers, all had been destroyed or scrapped by the early 1970s.

There were, of course, other large wartime aircraft projects, but for the most part these did not fare as well as the Globemaster. An exception was the Martin Mars, a flying boat transport derived from the XPB2M-1 patrol bomber prototype that first flew on July 3, 1942. The lack of suitable airfields in many parts of the world at the time made flying boats attractive, and although the Mars never went into production as a patrol aircraft, it did prove quite capable in the transport role. War's end cut the planned production of 20

Pratt & Whitney TF33-P-7 turbofan engine with its cowling uncovered. The C-141 was the first new military aircraft designed from the outset to use the new engine, although B-52Hs and some transport and electronic warfare versions of the C-135 were built with versions of the powerplant. (Chris Reed)

C-141 cargo hold, looking aft. The Starlifter had the same fuselage cross-section as the earlier C-130 Hercules, although the C-141 was considerably longer. (Air Mobility Command Museum)

C-141 "front office." The Sarlifter was notable in that it used vertical tape displays for many instruments in place of conventional round dials. (Air Mobility Command Museum)

JRM-1 versions by two-thirds; an additional aircraft was built to the refined JRM-2 standard, with earlier machines being reworked as JRM-3s. Capable of carrying 300 passengers or up to 20,000 lbs of cargo, the Mars was to be the largest USN flying boat to see squadron service. Released by the Navy in the mid-1950s, four Mars flew on as civilian waterbombers; a pair were subsequently written off, but the remaining two were still fighting fires in 1999.

As large as the Mars was, it was dwarfed by the Hughes H-4 Hercules. Nicknamed "Spruce Goose" because much of its construction was of non-strategic wood (birch rather than spruce), the H-4 originated as the HK-1 project of Henry Kaiser of Liberty Ship fame, who intended to build a "ship of the air." Free of runway constraints and invulnerable to submarine attack, the Hercules was designed to carry over 60 tons of cargo across the Pacific. Although not primarily intended for use in the passenger transport role, had such a version been built it would have been able to carry 700 men.

The H-4 was too late for the war, and like the C-5A two decades later, became something of a political nightmare for its creator, with Congressional critics and others accusing Howard Hughes of in effect stealing funds for an aircraft that could not fly. To refute this, Hughes himself took the H-4 on its first and only flight, taking the craft just barely airborne over Los Angeles Harbor on November 2, 1947. It was never flown again, and no further examples were built, although Hughes kept it in secluded storage for decades.

After Hughes died a recluse in 1976, there was actually some interest in refurbishing the aircraft as a testbed for U.S. Navy research into Wing In Ground Effect (WIG) vehicles. This was never carried out, and after years of display at Long Beach, the H-4 was dissambled and moved by sea to Oregon for rebuilding and eventual museum display.

Convair's entry into the field was the XC-99 freighter version of the B-36 Peacemaker, mating a new, larger fuselage to the B-36's engines, wings, and empennage. First flown on November 23, 1947, the XC-99 was the first aircraft capable of carrying 100,000 lbs of cargo; 400 troops could be accomodated, or 300 stretchers in the aeromedical evacuation role.

Although no further C-99s were built, the prototype did see much active service; in April 1949 a fifty-ton payload was carried aloft for a world record, and four years later it established an intercontinental cargo record of thirty tons on its first trip across the Atlantic. The largest piston-engined transport ever operated by the USAF flew on until 1957; preserved after a fashion at Kelly AFB, Texas, it was still in existence, in very poor condition, by early 1999.

And finally, Lockheed built a pair of Model 89 Constitution prototypes for the U.S. Navy, which designated them as XR60-1s. First flown in 1946 from Muroc AAF and powered by R4360s, the double-deck Constitutions were to have been followed by production models with either Wright Typhoon turboprops or improved R3350 radial engines. The USN never purchased any additional Constitutions, with the prototypes eventually finding their way into civilian colors. One survived, in derelict conditions, until the early 1970s.

Other, smaller wartime prototypes also introduced ideas that would later be used in large airlifter design. Budd's RB-1 Conestoga and the Curtiss C-76 Caravan both had high-mounted wings and high-set flight decks to maximize cargo compartment volume; the Conestoga had a rear door/ramp, and the C-76 had a hinged nose, which would ease loading considerably.

As would be the case with the C-5A program two decades later, builders of the large military transports of the late 1940s envisioned using these designs as springboards for commercial versions, anticipating a huge upsurge in civilian air travel. Douglas marketed the C-74 as the DC-7 (a designation that would later be used by

C-141B on a 1992 mission to support the USAF's Thunderbirds acrobatic team. This aircraft is finished in the AMC overall gray scheme, adopted after the European 1 scheme caused overheating during desert operations. (Chris Reed)

another aircraft), and Convair proposed a passenger derivative of the XC-99, as did Lockheed with the Constitution. Although commercial air travel did come into its own in the 1950s, none of these designs would ever take to the air, as airlines were well supplied with Boeing 377s, Lockheed Constellations, and Douglas DC-6s and DC-7s, as well as huge numbers of surplus DC-3s and DC-4s.

Despite the massive defense budget cuts that took place in the post-WWII/pre-Korea era, the development of new airlifters was still pursued. The first outsized transport to enter service with the newly independent USAF was Douglas' C-124 Globemaster II, first flown on November 27, 1949. Although using a new designation, this aircraft was actually an outgrowth of the C-74, with the YC-124A prototype being a reworked Globemaster I. An entirely new, much more voluminous fuselage was adapted to the existing tail and wings. A double-deck configuration for troop carrying would accomodate up to 200, or a slightly fewer number of patients and attendants if carrying out an aeromedical evacuation mission. Loading of large equipment and vehicles was accomplished by clamshell doors in the nose; to aid in cargo handling, inside the hold were a pair of eight-ton capacity overhead cranes.

The Globemaster II entered service quickly, with the first of 204 C-124A production models arriving in the spring of 1950. The YKC/YC-124B was a one-off YT34-powered model built to serve as the prototype for a tanker version; this did not go forward, but the later C-124C did, this model having wingtip heater pods and an APS-42 weather radar fitted in a nose "thimble," along with the more powerful R4360-63 engine. Altogether, 448 Globemasters were built by the time the line shut down in the spring of 1955.

The C-124 would be first blooded during the fighting in Korea; although capable of spanning the Pacific with a single refueling stop under some conditions, the Globemaster also operated into South Korea from Japan, with two squadrons of the 374th Troop Carrier Wing converting from C-54s in 1952. As would be the case in later operations, the USAF had to cope with forward airfields that were not suitable for such large aircraft; operating from Tachikawa, Japan, the C-124s were only able to fly from a handful of fields on the Korean peninsula, and even then could only carry up to 18 tons of payload.

There were some notable Globemaster missions during the war, including the supply of drop tanks that were urgently needed for F-86 sorties into "MiG Alley," and the record aeromedical airlift of 167 wounded to Japan in 1951. But there was also a major disaster involving a C-124, near the end of the war. On June 18, 1953, while attempting to take off from Tachikawa, a Globemaster crashed after an engine failed, and all 197 aboard were killed, making the crash the worst single aviation accident to that time, exceeding a December 20, 1952, Globemaster crash at Moses Lake, Washington, that killed 87. The crash led to a grounding of C-124s until repairs could be made; this could not have come at a worse time, given the situation in Korea.

The Globemaster II would be the centerpiece of USAF heavylift operations into the early 1960s. Post-Korea missions included supporting nuclear testing in the South Pacific, bringing in supplies for the building of the DEW line in northern Canada, helping with the deployment of Thor and Jupiter ballistic missiles in Europe, airdropping supplies over Antarctica, and supporting airlift operations to Lebanon, the Congo, and the Dominican Republic. Logistical missions included the ferrying of C-130 horizontal stabilizers and AGM-28 Hound Dog missiles.

The USAF was not the only service to make use of the C-124's capability to transport missiles, as Globemasters also helped with the deployment of Army MGM-18 Lacrosse units, as well as taking support equipment to Fort Sill during testing of Martin's Pershing I SSM.

Lockheed made use of the C-124's cavernous interior to transport, under heavy security, the partially disassembled airframes of

Made famous by a B-17, the name "Memphis Belle" has been applied to several other aircraft over the years, including a C-141B. (Paul Osborne)

Aerial refueling capability had not been fitted to the Sarlifter fleet on the assembly line, but the 1973 Nickel Grass airlift showed the necessity for airlifters to be so equipped, and refueling receptacles were fitted as part of the C-141B rebuild. (Chris Reed)

NC-141A 61-2779 Against the Wind started out life as the third Sarlifter built, ending up as the Advanced Radar Test Bed with a fighter-type nose. Seen here in June 1992, it was part of the then-new Air Material Command, assigned to the 4950th Test Wing at Wright- Patterson AFB, Ohio. (Chris Reed)

U-2 spyplanes from the Skunk Works plant at Burbank, California, to the secret "Ranch" testing facility at Groom Lake, Nevada, with the first such lift of the prototype "Article 341" using two Globemasters, one to carry the fuselage and the other the spyplane's wings. Moving the classified aircraft by air helped keep the program out of the public eye, as well as expediting the shipping process.

Globemasters assisted NASA by airlifting equipment for Mercury missions, and there was also C-124 support of the later Gemini program, such as the ferrying of the Gemini 3 capsule from the assembly plant to Cape Kennedy in January 1965.

A major role for the C-124 was the support of SAC's bomber force, which at the time was the country's sole means of strategic nuclear deterrence. Among the units so tasked in the early-to-mid 1950s was the 19th Logistics Support Squadron at Kelly AFB, Texas, which was equipped with 13 C-124As, including the prototype aircraft. Although assigned to Air Material Command, the 19th supported SAC. With the nuclear age less than a decade old, the USAF did not yet have an extensive stockpile of weapons, and at the time did not permanently station weapons overseas. C-124 flights would take the bomb assemblies to forward bases for varying periods of time before returning them to the U.S., thus keeping the Soviets unsure as to whether weapons were present at a base at any given time.

Carrying bombs abroad could involve three aircraft: one to carry the major assemblies; one to carry the plutonium "bird cages"; and another to transport the trigger mechanisms. As would be expected, security on the nuclear transport flights was high, with each mis-

Starlifter navigator's station. The AGAR 11 nameplate refers to a radio callsign commonly used by the 4950th, and later 418th Flight Test Wings. (Air Mobility Command Museum)

The NC-141As were unusual in that they wore the old Military Airlift Command gray/white scheme until retirement. 61-2775 is shown here at the Air Mobility Command Museum, Dover AFB, Delaware. (Air Mobility Command Museum)

sion carrying four Air Police armed with sidearms and automatic weapons. On the ground, the APs provided a protective detail for the aircraft, assisted with loading, and continued to monitor the weapons/materials while airborne.

There were several accidents during Globemaster nuclear missions. Although complete records of such mishaps still remain classified, it is known that a C-124 crew had to drop a pair of unarmed bombs off the eastern seaboard in late July 1957; engine failures had mandated the jettisoning if the transport was to stay aloft. The crew safely recovered the Globemaster at an airfield in New Jersey; it was disclosed in the press that cargo had been dropped, but the Air Force did not at the time admit that weapons were involved. Just under two years later, on July 6, 1959, another Globemaster carrying nuclear material was lost on takeoff from Barksdale AFB, Louisiana, and in the resulting blaze a weapon was destroyed. There was some contamination from the nuclear material, but no detonation. The last known C-124 nuclear accident took place on October 11, 1965, when a Globemaster burned out during a ground fire at Wright-Patterson AFB, near Dayton, Ohio, with only minor contamination resulting.

Aside from hauling nuclear weapons, C-124s also helped keep SAC's bombers ready for action, carrying burnt-out R4360 engines from B-36 Peacemakers back to the U.S. for overhaul. These massive powerplants were loaded up the front ramp by winch and tow vehicle.

NC-141A 61-2777 was used for testing the AN/ALQ-101 electronic countermeasures system for the B-1 bomber, and received an extensively modified tail for mounting antennas and other systems. (Paul Hart)

NC-141A 61-2777. (Paul Hart)

Globemasters also carried out their share of mercy missions. After flooding had wiped out telephone lines between the U.S. and Mexico, a C-124 brought in telephone cable, trucks, and personnel to repair the damage. This was no easy task, as the aircraft had to land on an asphalt runway that was far too short, and in any case, was just barely above the floodwaters itself. A preliminary defuelling stop had to be made on the U.S. side of the border to bring the weight down; the Globemaster's austere field capability was limited, the main gears having a single large wheel apiece.

In late December 1964, rain and melting snow caused major flooding in northern California and Oregon, killing dozens, destroying roads, railways, and bridges, and putting whole communities under water. President Johnson declared the region a disaster area, and a massive relief operation using fixed and rotary-wing aircraft was launched to rescue victims and supply isolated towns. In order to deal with the large number of aircraft operating the area, often in poor or no visibility over rugged terrain, the USAF deployed radar equipment to the base at Arcata Airport, California, aboard C-124s. Globemasters also hauled in fuel trailers to keep rescue aircraft operating.

Other C-124 mercy missions included flights in response to flooding in Tehran, Iran (August 1956), Somalia (November 1961), and Morocco (January 1963), as well as earthquakes in Morocco (February 1960), Iran (September 1962), Skopje, Yugoslavia (July 1963), and Sicily (January 1967). Refugee airlifts were also carried out, such as the removal of U.S. citizens from Arab countries upon the outbreak of the Suez Crisis in late October 1956.

As the 1960s dawned, the Cold War between the superpowers was threatening to turn hot in several places, especially Berlin. Tensions over the divided city heightened when Soviet Premeir Nikita Kruschnev had the Berlin Wall erected, putting his country and the U.S. on a collision course. NATO made preparations for a second Berlin Airlift, an operation in which the C-124's voluminous cargo capacity was expected to play a major part. A Warsaw Pact blockade did not in fact occur, but Globemasters did see use in Europe during the Berlin Crisis. To strengthen NATO air forces, the USAF in the fall of 1961 initiated *Operation Stair Step*, the mobilization and deployment of Air National Guard fighter units. While ANG F-84F and F-86H squadrons self-deployed to Europe, three units flying much more modern F-104s had their aircraft dissambled and shipped over in the cargo holds of C-124s.

Despite its age and rather ponderous speed, the C-124 would continue on in service even as the C-141A began to replace it. As late as 1968, Reserve squadrons were mobilized as part of the response to the Tet Offensive and rising tensions on the Korean peninsula; these aircraft saw use in the February 1968 deployment of a Brigade of the 82nd Airborne Division to South Vietnam. Also during that turbulent year, C-124s, along with C-141s, C-130s, and C-97s, assisted with airlifting forces for Operati*on Garden Plot*, the military effort to assist in restoring order to American cities in the wake of rioting set off by the assasination of Dr. Martin Luther King, Jr.

"Starlizard" in European 1 colors. (Air Mobility Command Museum)

Like the rest of the AMC fleet, the surviving C-141Bs have been refinished in overall gray. This is likely to be the Sarlifter's last paint scheme, as all are to be retired by 2006. (Paul Hart)

Reserve squadrons continued to fly the type on missions well into the 1970s, such as the airlift of the prototype YA-10 from Farmingdale, New York, to Edwards AFB by two C-124s. The final Globemasters were not retired until 1974, a quarter-century after the YC-124A had first flown. In 1984, C-124 52-1000, following years of storage at Aberdeen, Maryland, flew from Dover AFB to California for display at Travis AFB.

Douglas was to produce a third heavy airlifter, the C-133 Cargomaster, which first flew on April 23, 1956. This was an entirely new design owing nothing to the C-124, although it did resemble in general layout an outsized version of Lockheed's C-130 Hercules. The circular cross section fuselage was some 157 feet long; 90 feet of this comprised the pressurized cargo hold, loadable through a two-part rear door. Like the Hercules, the Cargomaster had its main landing gear housed in fuselage fairings to maximize the available space within the hold.

The high-mounted wing had fitted to it four Pratt & Whitney T34 turboprops, giving the Cargomaster a top speed of nearly 360 mph. The Cargomaster would be the first service aircraft to be powered by the T34; early air testing of the engine had been conducted aboard a B-17 testbed, while service trials were flown using a pair each of YC-97 and YC-121 testbeds.

Entering service in 1957, the Cargomaster was to equip two units of MATS, the 1607th ATW at Dover, Delaware, and the 1501st ATW at Travis, California. In an era when the jet fighters and bombers of TAC and SAC were routinely setting new speed, altitude, and distance records, MATS used the Cargomaster to set some records of its own; on December 16, 1958, a C-133A flying from Dover established a record by taking a 117,900 lb payload to 10,000 feet.

The production run was finished with the procurement of 15 C-133B models with uprated -9W engines, along with clamshell doors. These aircraft saw use in transporting ICBMs from factories to airfields near their intended launch complexes, the Cargomaster being able to carry even the large Titan and Atlas missiles; first flight of a B-model took place on October 31, 1959. Flights were also made carrying missiles to Cape Canaveral for test launches, a journey that would take days by surface transportation being covered in mere hours by air.

Ironically, the Cargomaster was coming into service just as work was beginning on a substantially smaller ICBM, the Boeing LGM-30 Minuteman. Minuteman deployment would outstrip that of all other U.S. ICBMs combined, and supporting this effort would be a major mission for the C-133. LGM-30 logistical support centered at the Utah Air Logistics Center, from where assembled missiles, minus their warheads, were shipped to operating units. The missiles were mounted on a structural strongback and encased in a travel container; upon arriving at an airbase they were then transferred to a Transport-Erector vehicle for the road trip to their silos. The Transporter-Erector itself could be moved by air.

NASA also made use of the C-133's ability to airlift large missiles and spacecraft, including several flights to take Titan II launch vehicles to Florida for use in the Gemini program. Cargomasters supported the space agency in other ways, as well. For example, on May 3, 1961, a C-133 flying over El Centro, California, dropped the third "boilerplate" version of the Apollo Command Module to verify the unmanned testbed's parachute landing system.

There were also several proposed space support roles for the C-133 that were never implemented. For example, the problem of moving the Saturn booster's S-IVB upper stage to the Cape was particularly thorny, as it was far too large to fit inside conventional aircraft. Ultimately, the distinctive Supper Guppy aircraft were created for this mission, but an alternative plan, discussed at least as early as the fall of 1961, would have used a Cargomaster with fittings to carry the stage externally on its back. And prior to the cancellation of the X-20 Dyna-Soar program, the C-133 had been considered as a launch vehicle for early air-drops of the Boeing-built space glider.

Despite its capabilities, the C-133 did encounter its share of problems. Between June 1961 and January 1965 eight aircraft, or

Photo of C-141B 64-0626, Copilot's position. (Air Mobility Command)

66-0209 of the 437th MAW, March 25, 1983. (Masanori Ogawa)

65-0234 of the 60th AW, still wearing the old gray/white scheme on September 5, 1993. (Masanori Ogawa)

nearly one-fifth of the inventory, were lost in accidents. With the exception of one aircraft lost in a ramp fire, all of these losses occurred in the air during landing or climbout, with no warning. This led to a flight test program to better understand the Cargomaster's stall characteristics, propellor modifications, and the fitting of flight data recorders.

Though troubled, the Cargomaster did provide the USAF with an outsized airlift capability that would not be exceeded until the advent of the C-5A. Much of the Army's mobile equipment of the era could be carried, a capability that would be put to good use on airlift missions to Southeast Asia. ARVN M41 tanks were brought into Da Nang during the Buddhist revolt; more than half a decade later, C-5s would carry out similiar armor airlift missions to the base. Aircraft brought into the theater in the hold of C-133s included Sikorsky HH-3Cs in 1965, and the first Bell AH-1 HueyCobra gunships in the late summer of 1967. Cargomasters also transported in OV-10 Broncos and Air Cushion Vehicles.

Retired from military service in the early 1970s, the Cargomaster fleet was, for the most part, quickly broken up for scrap, but a few survived to receive civil N-numbers, and by early 1999 a solitary example was still maintained in Alaska.

The Cargomaster was actually supposed to be the smaller of two giant Douglas turboprops, with the other being the XC-132. This colossus would have been capable of lifting more than fifty ton payloads, with the aircraft grossing at 500,000 lbs. The double-deck fuselage would be mated to high-mounted swept wings and powered by four Pratt & Whitney T57s, each rated at 15,000 hp. The projected performance was impressive, with a top speed of 460 mph and a range with 50,000 lbs payload of 3,500 miles.

A tanker KC-132 was considered, although the arrival of the faster, pure-jet KC-135A would have probably kept any production plans from being carried out. And like the C-5 of a decade later, the XC-132's sheer size made it a potential carrier for standoff weapons, in this case a pair of huge Northrop Snark cruise missiles.

The XC-132 mockup was first revealed to the public in February 1957, but by that time the program was already in trouble. Attempting to forestall the "military industrial complex" he feared, President Eisenhower was cutting the military budget, and faced

63-8085 of the 63rd MAW, April 11, 1987. (Masanori Ogawa)

66-0128 of the 63rd, June 6, 1983. (Masanori Ogawa)

67-0166 of the 433rd, visiting Yokota, March 15, 1989. (Masanori Ogawa)

65-0239, 60th MAW, March 13, 1992. (Masanori Ogawa)

with fewer funds, the USAF elected in March of that year to cancel the XC-132 rather than a combat aircraft.

Another never-built heavy airlifter of the late 1950s, the Martin SeaMistress, was in many ways a hybrid design, combining the company's decades of experience in seaplane design with modern turbine powerplants. The SeaMistress would be based broadly on the company's earlier P6M SeaMaster, but in greatly scaled-up form. The much larger fuselage would have had a visor nose to ease unloading of cargo onto barges or docks. Power would be provided by six to ten Pratt & Whitney J75 turbojets built into the wing, but developed versions were intended with more advanced turbofan engines.

The SeaMistress' potential benefits were numerous; it could operate without the need for long hard-surface runways, being able to fly to and from coastal areas and even rivers and lakes. Payload capacity would exceed a quarter-million pounds for short-range missions, and the aircraft could land at sea as needed to take on fuel from ships.

SeaMistress never got beyond the "paper airplane" stage—the general low priority put on specialized airlifters at the time was one factor behind this, but the fact that the U.S. Navy might have attempted to control a large part of the strategic airlift mission by operating the SeaMistress quite possibly caused alarm within the Air Force.

Enter the C-141

By the dawn of the 1960s, the Military Air Transport Service was equipped with a large number of propeller-driven transports, on paper a major force. However, these aircraft were ill-suited to the changing face of warfare. Although the potential for a strategic nuclear exchange remained, there was also now concern about regional conflicts between U.S. and Soviet-backed proxies that could tip the balance of power, or even precipitate direct combat between the superpowers. Faced with countering global contingencies, the U.S. armed forces were to place more emphasis on rapid mobility; the Army in particular was concerned with quickly moving divi-

63-8088, 60th MAW, November 29, 1981. (Masanori Ogawa)

65-0241 of the 62nd, still in European "lizard" colors on August 17, 1999. (Masanori Ogawa)

sions in the United States directly to Western Europe in case of a crisis with the Soviets and Warsaw Pact. Prepositioning supplies in depots and aboard ships near potential crisis zones was only a partial answer, as this would mean that additional material and vehicles would have to be bought for U.S.-based troops to train with.

As shown in early 1960s mobility exercises and subsequent airlift operations to Southeast Asia, the C-124 was simply too slow for efficient long-range operations, as such missions demanded too many flying hours. The limited number of C-133s were beset with structural problems that restricted their capabilities; clearly, a faster aircraft was needed. Speed was not the only consideration, as performance was needed to be matched by a design that was optimized from the start for the military mission. While the C-133 had been built for the airlift role, the bulk of the MATS fleet consisted of such types as the C-97 Stratofreighter, C-118 Liftmaster, and C-121 that were little different from their commercial Boeing 377, DC-1, and Constellation counterparts, and thus had little capability to haul outsized items.

As a short-term fix, MATS did buy the turbofan-powered C-135B Stratolifter version of the KC-135A tanker, as well as the extended range C-130E Hercules, but these were only interim steps. The Hercules did not have the speed performance required, while the much faster C-135B had its own shortcomings. Like the types it helped replace, the Stratolifter was closer to a commercial aircraft than a purpose-built airlifter; loading could only be accomplished through a fuselage cargo door high off the ground, and the low-mounted wings impeded ground loading operations. Additionally, the Stratolifter had no capability to drop paratroops. These factors did not hinder the design's overwhelming success as a tanker, but they did help restrict C-135B purchases to 30 aircraft, preceded by 17 C-135As with the original J57 turbojet engines.

Even with its limitations, the Stratolifter did much to validate the idea of a jet-powered airlifter. In January 1962, the mobility exercise *Long Thrust* was held, aimed at demonstrating the capability to rapidly move three Army battle groups from McChord AFB, Washington, to West Germany. Of the approximately 100 aircraft involved in the exercise, only seven were C-135s, but these performed all out of proportion to their small number, taking troops directly to Germany over the polar route, a distance of over 5,000 miles. Much of the group's equipment had been positioned in Germany beforehand, but *Long Thrust* nonetheless showed what the new breed of airlifters were potentially capable of. Following the end of maneuvers, one battle group of 1,500 men was brought back to the U.S. by 21 C-135 missions. C-135s also took part in the reinforcement of Guantanamo Bay, Cuba, during the Missile Crisis of 1962 (aircraft 62-4136 was lost in a crash there on September 29), the airlift of arms and equipment to India following the Chinese invasion of 1962, and the *Operation Biglift* movement of the 2nd Armored Division from the U.S. to Germany.

64-0643, 62nd AW, January 16, 1999. (Masanori Ogawa)

65-0242, 60th AMW, June 14, 1997. (Masanori Ogawa)

The pioneering Stratolifter was only a stopgap, as an all-new type designed for the role was required, a need that was to be met by the airlifter team at Lockheed Georgia, putting the company on the road that would eventually lead to the C-5A. Expanding U.S. airlift capabilities was one of the high priority items in the early military planning of the Kennedy Administration, and shortly after JFK's inauguration in January 1961, Lockheed, Boeing, Convair, and Douglas had presented their plans for the proposed CX-1 jet airlifter. Also known as SS-476L, the CX-1 was to meet the Specific Operation Requirement 182 for a jet transport able to haul 70,000 lb payloads over short distances, 50,000 lbs across the Atlantic, or 20,000 lbs across the Pacific.

Lockheed was able to make use of its experience with the ongoing C-130 program, as well as late 1950s efforts to build a larger version. The GL-500 Super Hercules was to be a more powerful, stretched model. Most planning centered around a version powered by four of the new Allison T61 turboprops, but British Rolls-Royce Tyne engines were also considered. The Super Hercules was never built, but it did lay much of the groundwork for Lockheed's CX-1 entry, which essentially used a stretched Hercules fuselage mounted to high-mounted 25 degree swept wings and a T-tail. The

67-0009 of the 63rd, April 5, 1987. (Masanori Ogawa)

L-300 Starlifter design was chosen in March 1961 for procurement as the C-141A. Just as its Hercules predecessor had set the design standard for turboprop airlifters, so did the Starlifter establish a basic configuration for jet types.

A new advance in engine design was pivotal in the development of the CX-1 program, and helped pave the way for the later C-5. Whereas early pure turbojet engines were notorious gas guzzlers, by the early 1960s engine manufacturers were offering the first generation of more powerful and fuel-efficient turbofan powerplants, which added a forward fan stage to increase airflow around the core of the engine. By March 1961, when Lockheed's airframe design was chosen, Pratt & Whitney's JT3D had been picked to power the new aircraft, beating out a General Electric competitor. Developed from the JT3/J57 line, the JT3D-3 engine, rated at 18,000 lb thrust, was replaced before the first C-141 flight with the more powerful JT3D-8A, which would be given the military designation TF33-P-7. Unlike some JT3D configurations, the -8A had nacelles that carried the exhaust from the forward fan stage all the way aft, resulting in a much sleeker installation. Fitting of a more powerful TF33 variant originally intended for the never-built Boeing 707-820 was considered for later C-141s, but was never carried out.

Like its smaller brother the Hercules, the Starlifter was designed with paratroop operations in mind; indeed, it was the first jet transport employed in that role. At the rear of the cabin, paratroop doors were fitted on both sides of the fuselage, and 123 side-facing seats could be installed. When transporting more conventional passengers, up to 154 aft-facing seats were fitted. The aft ramp could be opened in flight for equipment drops, and on July 12, 1965, a record cargo airdrop was achieved over El Centro, California, when a C-141 dropped seven pallets weighing a total of 70,195 lbs in under half a minute.

Rollout of the first aircraft took place on August 22, 1963, several days ahead of schedule. As evidence of the program's political importance, President Kennedy personally signaled for the hangar doors to be opened from Washington, where he was watching the event from a closed-circuit TV broadcast. None could have guessed at the time that the initial Starlifter, 61-2775, would outlast Kennedy and several of his successors. Tinker AFB, Oklahoma, was the first to receive the Starlifter, in October 1964.

Travis AFB, California, began its three-decade history with the Starlifter on April 23, 1965, when the first aircraft arrived to equip the 1501st Air Transport Wing. This Starlifter proved to be another long-lived example, as it returned to its first home station some thirty years later to finish out its service life. The 1501st was replaced by the 60th Military Airlift Wing on January 8, 1966.

The 1608th ATW at Charleston, South Carolina, received its first aircraft, aptly named City of Charleston, on August 14, 1965. On January 8, 1966, the unit became the 437th MAW and traded in its E-model C-130s the following year, retiring its obsolete Globemaster IIs two years after that.

The C-141 came into service as the conflict in Vietnam was being joined in earnest, and the Starlifter was to prove pivotal in U.S. logistics operations to support the fighting. Piston-engined airlifters were still common sights in the theater, but the C-141 soon became the backbone of the strategic airlift effort. Indeed, the first Starlifter mission to carry troops operationally was related to the situation in Vietnam, with the 44th transporting nearly 100 USAF personnel to the Philippines on July 21, 1965. A week earlier, a cargo mission had been flown to Yokota AB, Japan.

The Starlifter had not been in service long before its first major wartime test took place. As American involvement grew, more and more units received their deployment orders, and in December 1965 the 25th Infantry Division at Schofeld Barracks, Hawaii, learned that it was their turn to head for Southeast Asia. Although most of the division would be moved by sealift in the spring of 1966, the 3rd Brigade, with nearly 3,000 troops and 5,000 tons of equipment, would be moved by Starlifters, C-124s, and C-133s to Pleiku, in the central highlands of South Vietnam. The lift, codenamed *Operation Bluelight*, was accomplished between December 23, 1965, and January 23, 1966.

63-8086 of the 62nd, January 16, 1994. (Masanori Ogawa)

As big as *Bluelight* was, it was soon to be surpassed by an even larger lift. Some elements of the 101st Airborne Division "Screamin' Eagles" had been in Vietnam since 1965, but as part of the build-up it was decided in the fall of 1969 to deploy the balance of the division from Fort Campbell, Kentucky. *Operation Eagle Thrust*, as the lift was codenamed, was to predominately be a C-141 operation, with 367 Starlifter missions being flown.

The distance between Campbell and the airbase at Bien Hoa, South Vietnam, varied between 9,700 and 10,500 miles, depending upon which route was taken over the Pacific. Despite having to fly nearly halfway around the world, the Starlifters covered the distance in a fraction of the time that would have been necessary with propeller aircraft.

On return flights to the U.S., Starlifters served as aeromedical evacuation aircraft, a role that would become one of the type's most visible missions in the years to come. The capability to rapidly return casualties from Southeast Asia to facilities in the U.S. was due in large part to the C-141A. Peacetime evacuation flights were mounted for victims of bombings, airliner crashes, and other disasters.

Following the *Linebacker II* air campaign of December 1972, stalled peace talks in Paris between the U.S., South Vietnam, the Viet Cong, and North Vietnam resumed, and in January 1973 a cease-fire agreement was finally reached, ending the U.S. combat role in Vietnam. Of major importance to many in the U.S. was the resolution of the POW/MIA issue, as some American servicemen had been missing or held captive since 1964.

Although thousands would remain missing in Southeast Asia even a quarter-century after the war's end, 1973 would see the return of over 500 men from captivity in both North and South Vietnam as part of *Operation Homecoming*. Retrieving former captives from the north would put Starlifter crews in the unusual situation of operating into the Gia Lam airport in Hanoi, which had been the target of U.S. bombs just several months before.

The first former prisoners out of North Vietnam departed on February 12, 1973, on C-141s bound for a stop at Clark AB, the Philippines, before returning to the United States. The first Starlifter into Gia Lam, 66-0177, would still be in service as a C-141B twenty-five years later, having been christened *Hanoi Taxi* and serving as a memorial to the POW return operation.

Homecoming would not be the only post-cease fire airlift activity, as the treaty also stipulated the withdrawal of the nearly 24,000 U.S. military personnel in South Vietnam. C-141s participated in this lift, *Operation Count Down*, which also entailed the withdrawal of South Korean forces in Vietnam, which numbered some 37,000.

Little more than two years later, Starlifters would be back in South Vietnam, helping to bring out those fleeing the final advance of the NVA. And several weeks after the fall of Saigon, a C-141A was used to take Marines from Okinawa to Thailand to take part in the operation to recapture the SS *Mayaguez*.

One of the most somber tasks given the Starlifter was the return of deceased American servicemen (and occasionally, civilians) from overseas. For years after the end of the Vietnam conflict, C-141s would periodically make flights back to Southeast Asia to pick up recovered remains of missing servicemen, returning them with military honors to the U.S. for identification and proper burial. The reparation mission had passed to C-17s by the end of the 1990s.

Certainly, some of the more unpleasant such missions carried out by C-141 crews were the casualty removal flights out of Guyana in November 1978. Alarmed at reports coming out of Jonestown, the People's Temple community of expatriate Americans in Guyana, Congressman Leo Ryan led a fact-finding mission there, only to be killed along with several others at a nearby airstrip. This was only the beginning of the Jonestown slaughter, as a mass suicide in the community then ensued.

Confronted with hundreds of bodies lying exposed in a tropical climate, the government of Guyana asked the U.S. on November 21 to remove the remains, leading to a U.S. military deployment to the country. As the distance between Jonestown and the Starlifter-capable airfield at Georgetown was over 100 miles, it was necessary to bring in heavy-lift helicopters to transfer the remains, specifically HH-53 "Jolly Green Giants" flown directly into Guyana,

66-0159, 62nd AW, April 18, 1994. (Masanori Ogawa)

having refueled from HC-130s along the way. Smaller UH-1 helos were also flown and airlifted in to assist in the operation, and C-130s were deployed from Panama.

The U.S. personnel deployed to Jonestown had the grisly task of loading the remains into body bags on site and then flying them to Georgetown, where they were transferred into cases and put aboard the C-141s. Despite being enclosed in the transfer cases, the decaying bodies put off such a stench that aircrews often had to use oxygen to withstand the smell on the long flights to Dover.

The casualty toll, which at first report had numbered less than one hundred, kept climbing, and by the time the airlift had ended on November 26, over 900 bodies had been brought to Dover for processing and temporary storage in an alert hangar.

Other Starlifter casualty flights included the return of the Marine Colonel William Higgins after his death while a captive in Lebanon, and the remains of a U.S. Army crew that perished in the crash of their RC-7 in Columbia during July 1999.

Special Purpose Models

From the outset of the CX-1/SS-476L program, it was intended that the new aircraft would, as part of its testing, be certified by the FAA, facilitating the entry into service of a commercial variant. Lockheed had high hopes for the concept, as a larger complement to projected civilian C-130 versions, and as a competitor to 707 and DC-8 freighter variants.

Commercial Starlifters, to be given the designation L-300, would likely have had their engines slightly derated, but even so could have carried 96,000 lb payloads. FAA certification of the Starlifter was completed on January 20, 1965.

There was also interest in a stretched model, the L-300B, which would have had a fuselage some 23 feet longer than that of the C-141A, this being achieved by inserting a 15 foot plug forward of the wing and an 8 foot section aft, increasing the cube volume to 13,750 cubic feet. In 1964, a tentative order for four L-300Bs was placed by Slick Airways. The Slick aircraft were never to actually be built, but the L-300B concept would resurface in military form more than a decade later as the C-141B.

66-0207 of the 305th AMW at McGuire, seen here at Yokota, August 9, 1997. (Masanori Ogawa)

Despite commercial sales having been a priority from the start, the basic C-141A/L-300 did not offer enough advantages over 707 and DC-8 freighters to merit orders, while the more capable L-300B was deemed to be different enough to warrant an expensive and time-consuming new test program. However, although production of the Starlifter for the USAF ended with the 284th aircraft, Lockheed did keep the line open to build a single additional aircraft, intended to demonstrate the type to potential commercial customers.

In spite of Lockheed's best efforts, the L-300 did not win enough firm commercial interest for the company to keep the line open. The prototype did not languish long, however, as NASA purchased the aircraft for use as an infrared observatory. IR observation of planets and galaxies offered scientists new ways to map and analyze celestial phenomena, but making such studies from the earth's surface was impractical, as water vapor in the atmosphere absorbs many infrared wavelengths. Suborbital rockets and telescope-equipped balloons were only partial solutions; what was needed was an aircraft able to take a sizable telescope up above 45,000 feet and keep it there long enough to carry out meaningful observations. The Ames Center had by the early 1970s operated a Learjet equipped with a ten-inch telescope, but a larger platform was needed, a need that the L-300 met admirably, and thus the one-off aircraft was taken on charge as NASA 714.

The centerpiece of 714's capabilities was a three-foot diameter Cassegrain telescope. The telescope compartment was isolated from the rest of the cabin by bulkheads, through which a mirror diverted the IR radiation, thus allowing crewmembers to work without oxygen masks and flight suits.

Modification of the L-300 to take the unit was carried out in 1973 by Lockheed Aircraft Systems, the telescope opening being on the port fuselage just forward of the wing fairing. In May 1975 the aircraft was formally dedicated, and named after noted astronomer Gerald Kuiper, subsequently being known as the Kuiper Airborne Observatory.

During its two decades of operation, the KAO gave astronomers an invaluable tool to research the universe, and added much to the understanding of earth's own solar system. KAO observations confirmed that Pluto had an atmosphere, and that those of the giant planets Jupiter, Saturn, and Neptune contained water. Orbiting rings were discovered around Uranus, and in 1994 missions were flown from Melbourne, Australia, to record the impact of the Shoemaker-Levy 9 comet fragments into Jupiter. The KAO also looked at the first reentry of the space shuttle Columbia to obtain data on the orbitor's interaction with the atmosphere.

The KAO has been deactivated, principally as a measure to free up funds for the SOFIA project, which will mount a 2.5-meter telescope in the aft fuselage of a converted Boeing 747.

Hopes for at least a limited number of Starlifter export sales were raised during the production run; France in particular could have used a small number to support nuclear testing in the South Pacific and its other overseas commitments, while West Germany and Canada were also potential customers. No foreign sales would be made, with France relying on its C-135F Stratotankers, while Canada and West Germany both bought Boeing 707s.

There were also plans to fit a C-141A wing on a version of the British Shorts Belfast turboprop freighter. Resembling in many ways a larger C-130, the Belfast would have also received a T-tail, swinging nose, and Rolls-Royce turbofans. Although never to be built, the Jet Belfast would have in some ways been superior to the Starlifter, having a fifty-ton payload capacity and a wider cross-section.

A small number of modified C-141As took over the ICBM transport role of the C-133B; thanks to strengthened floors, these were able to carry a Minuteman missile in its cargo container, although the refitted aircraft had a slightly lower g-rating than the rest of the fleet.

Around the time that the C-141A program was winding down, the type was considered as a platform for the new Airborne Warning And Control System (AWACS) that the USAF was funding to replace the EC-121. Preliminary concepts showed a conventional rotodome mounted dorsally aft of the wing, mounted on four support pylons. A fairing atop the forward fuselage was apparently an aerial refueling receiver and its associated plumbing. Ultimately, a version of the Boeing 707 beat out both the C-141A and a DC-8 AWACS derivative.

C-141B

As effective as the Starlifter proved to be, it was nonetheless hampered by its restricted fuselage cross section. While the C-5 was available for hauling outsized loads, the Starlifter force had airlift capacity that was going to waste, as the aircraft's weightlifting power could not be used to full advantage with the volume-limited airframe. Nothing could be done about widening the -141's fuselage, but as the L-300B concept and the L-100-20/30 programs had shown, stretching it would be entirely feasible. Adding 23 feet 4 inches to the aircraft's length via plugs forward and aft of the wings would increase the aircraft's clear cube volume to 11,399 cubic feet, an increase of 6,119 cubic feet over the C-141A. The stretched aircraft, to be designated C-141B, would now be able to haul thirteen 463L cargo pallets to the C-141A's ten.

Another major structural change would be the addition of a refueling receptacle to extend the aircraft's range, and thus cut dependence on intermediate bases that might not be available in times of crisis. The receptacle fairing was fitted just aft of the flight deck, with the fuel line running along the spine to enter the airframe at the wing/fuselage junction.

The aircraft chosen as the prototype was 66-0186 from Charleston, which was flown to Marietta, and rolled in its new form as the YC-141B in early January 1977. Lockheed's Frank Hodden and USAF Major Herb Klein conducted a successful first test flight on March 24th of that year.

Arriving at Marietta in February 1979, the first Starlifter to undergo a "production" refit was rolled out on July 18 of that year, and was delivered back to the USAF on December 4; the 60th MAW would be the first unit to be equipped, in the spring of 1980. The stretch program ended on June 29, 1982, when the 270th aircraft was redelivered.

Evidence of the B-model Starlifter's capability was shown in October 1980 during Operation REFORGER (Return of Forces To Germany), when C-141Bs flew to Europe and back nonstop with tanker support. Other long-range flights were carried out, with a C-141B reaching Dharan, Saudi Arabia, from McGuire AFB without stopping on March 30, 1981. Refueling three times from KC-135s, the Starlifter carried 67,000 lbs of cargo; this flight helped demonstrate the American ability to rapidly intervene in the Persian Gulf region at a time when tensions there were high over the Iran-Iraq War and the Soviet conflict in Afghanistan.

The Starlifter's new ability to take on fuel in the air also opened up a new mission for the type. C-141s had first been flown to Antarctica when an A-model landed at McMurdo Sound on November 14, 1966, becoming the first all-jet aircraft to operate on that continent. C-141As could only operate into McMurdo during the Antarctic summer, but B-models with tanker support could make nonstop round trip airdrop missions from Christchurch, New Zealand, even in the dead of winter.

Many flights have taken Starlifter crews to unusual locales, some of them having been off-limit previously. In February 1972 the 62nd MAW sent C-141A 66-0141 to China, carrying support equipment for the groundbreaking visit by President Richard Nixon that opened up relations with the government in Peking.

This would not be the only time that Starlifters from McChord would venture into China; in January 1997 a crew from the Air Force Reserve's 728 AS/446 AW at the base took a C-141B to the Chinese capitol (since renamed Beijing) to pick up the retrieved

Not all retired Sarlifters have been stored at the "Boneyard"; GC-141B 66-0189 served as a ground trainer at Sheppard AFB, Texas, with Air Education and Training Command markings. (Hans Heijdentrijk)

remains of a B-24J crew that had been lost in August 1944. And in August 1998 another McChord Starlifter arrived in the city of Changsa carrying relief supplies for flood victims.

The Starlifter was still MAC's most numerous strategic airlifter at the time of the 1990 Iraqi invasion of Kuwait, and virtually the entire force would be dedicated to the resulting Desert Shield airlift, helping to form the "air bridge" between the U.S. and the Persian Gulf. A C-141B had actually been on hand in the region even as the crisis began, supporting the Ivory Justice deployment of two KC-135s to the Gulf.

A C-141B carrying Airlift Control equipment and personnel was the first U.S. aircraft into Saudia Arabia in August 1990; in the following weeks and months a veritable flood of MAC transports would cycle in and out of the desert kingdom and many of its gulf neighbors. So critical was the need for airlift that work on aircraft undergoing depot maintenance was accelerated, and several Starlifters flew for a time without paint.

Getting the weapons, vehicles, equipment, and personnel to the Gulf was only part of the logistical problem. For years, critics of the American military had charged that U.S. systems were too advanced and unreliable, that they required far too much maintenance and spare parts support to be effective in actual combat. To counter this, on October 30, 1990, MAC put the Desert Express into operation. This was a daily C-141 shuttle running from Charleston AFB, South Carolina, to and ending in Saudia Arabia. *Desert Express* supplied deployed units with the so-called MICAP (Mission Capable) items that were needed to keep systems running. Material that regularly would have taken weeks to ship were delivered in days. Together with the *European Express*, *Desert Express* helped U.S. forces disprove the naysayers, with many weapons systems actually having readiness levels better than those achieved in peacetime.

Ironically, in a land rich in oil, U.S. forces for a time also had to bring in fuel to support some operations; the 9th SRW's U-2 operation in particular relied on Starlifter "bladder birds" to maintain its ability to generate sorties.

As the *Desert Shield* defensive deployment gave way to *Desert Storm* in January 1991, Starlifters continued to support Coalition forces, airlifting in munitions ranging from M117 gravity bombs for B-52 operations, to the GBU-28 *"Deep Throat"* bunker buster weapons, created in a matter of weeks and rushed to the theater aboard a C-141.

As would be expected, the strain of such intense flying soon took its toll on the Starlifter force, the youngest of which was twenty-two years old at the beginning of Desert Shield. Well before the war, in the late summer of 1989, cracks had been found in the inner/outer wing joints of a C-141, and further inspections found the problem to be widespread. Cracks were later found in other parts of the joints, as well, and by 1990 the fleet was operating under restrictions, including a 51,000 lb payload limit and curtailment of low-level operations.

To allow the USAF to keep 178 C-141s flying into the 21st Century, Lockheed proposed a major rebuild program, encompassing the aircraft receiving the Pacer Center retrofit, as well as 58 others. The company's plan involved replacing the aircraft's wing structure, replacement or reinforcement of other parts of the airframe structure, and a rehabilitation of some systems. Ultimately, it was decided to gradually retire the Starlifter rather than embark on a major refit program.

Bosnia-Herzegovina

With the experience of Desert Storm still fresh in its memory, the U.S. was starting to become involved in a very different type of conflict by the summer of 1992, this time in the former Yugoslavia. Bereft of the unifying force of the country's late founder, Josef Tito, Yugoslavia had broken apart in an orgy of civil war that would lead to a humanitarian crisis the like of which had not been seen in Europe for decades. The fighting in Bosnia-Herzegovina was particularly brutal, leading to a UN operation to airlift in food and relief supplies to thousands of starving civilians.

Although most of the USAF missions were mounted by C-130s operating from Rhein-Main, Germany, Starlifters also took part. As the airport at Sarajevo was supposedly a neutral area that was not to be shelled or otherwise attacked—this was recognized more in the breach than the observance—the USAF and other air arms fitted lightweight armor to the cockpits of their C-130s to protect the crews against small-arms fire; it is believed that some C-141s received this treatment, as well.

Special Operations Starlifters

Although the USAF operates a substantial force of MC-130 *Combat Talon* I/II aircraft, it has also relied on a small number of modified C-141Bs to carry out long-range special operations transport missions. Termed the Special Operations Low Level mission, this was carried out for years by the 16th Airlift Squadron at Charleston AFB.

The uprated SOLL-II aircraft can be distinguished from standard Starlifters by several "lumps and bumps," including undernose "pimples," which is actually the turret that provides the flightcrew

66-0228, another of Sheppard's GC-141Bs. (Hans Heijdentrijk)

with infrared imagery for night operations. Additionally, there are ECM blisters on either side of the nose. In order to provide such a large aircraft with some degree of self-defense capability, the type is fitted with the AN/ALQ-172 jamming system. Warning of radar-guided AAA and missiles is provided by the AN/ALR-69 RWR; active defense against SAMs and AAMs is handled by ALE-47 chaff/flare dispensers, which are cued by the AAR-44 Missile Approach Warning System.

As the 437th converted from the Starlifter to the C-17, the 16th lost the SOLL-II mission, which was being transferred in 1999 to McGuire AFB. This is likely to be the final station for the aircraft, as they too were scheduled to be replaced by Globemasters, in this case 15 modified aircraft ordered in the FY2000 budget.

Although airborne alert of nuclear-armed bombers ended in the late 1960s, aircraft with weapons aboard still take to the air, in this case C-141Bs tasked with the Primary Nuclear Airlift Force mission. As did their C-124 predecessors of nearly half a century earlier, PNAF transports take new or refurbished weapons into the field for deployment, and return them for modification, storage, or dismantling. This latter role has gotten the most use in recent years, as the U.S. has retired thousands of weapons after the end of the Cold War, ranging from gravity bombs in western Europe and South Korea to W69 warheads on SRAM missiles and W56s from retired Minuteman II ICBMs.

Obviously, PNAF aircraft rate the highest security, using isolated "hot cargo" parking spaces for loading/unloading under tight guard, with other aircraft being forbidden to overfly them.

C-141C

By the late 1990s, in the twilight of its career, the Starlifter was to undergo one final refit program. Although all-active duty aircraft were to be taken out of service by 2003, ANG and Reserve examples were scheduled to remain in service for an additional three years beyond that, meaning that some obsolete and unmaintainable systems would have to be replaced in the meantime. A total of 63 relatively low-time C-141Bs were put through the C-141C refit, which added a new flight control system, glass cockpit displays, chaff/flare dispensers, and SATCOM.

The USAF Reserve's 452nd AMW at March ARB, California, was the first unit to re-equip with the C-141C, accepting its first rebuilt Starlifter on October 31, 1997. All conversions were completed by August 1999, and with the conversion program ended and the NC-141As retired, Edwards AFB finally shut down the Starlifter flight test program in late July 1999.

NC-141A

Less well-known than the fleet of often strangely-modified NKC-135s, the small force of four NC-141As belonging to Systems Command (later Material Command) carried out their share of exotic testbed work over the years. Consisting of the first, third, and fourth airframes off the line, the NC-141As were the only surviving Starlifters not to become C-141Bs, and wore the old gray/white paint scheme until retirement. Flown for many years by the 4950th Test Wing at Wright-Patterson AFB, Ohio, the testbed Starlifters finished out their careers with the 418 TW at Edwards AFB, California.

61-2775, the very first Starlifter, was active in the test role well into the 1990s. During the fall of 1995, -2775 was again used as a transport to support NASA's STRAT (Stratospheric Tracers of Atmospheric Transit) program, wherein a NASA ER-2 conducted upper atmosphere sampling to determine the effect of aircraft exhaust on the ozone layer. The NC-141A ferried equipment and personnel from the NASA Ames Center to NAS Barber's Point, where the STRAT flights were staged from.

61-2775's final task prior to retirement was also one of the most unusual carried out by any of the NC-141s. In order to gain some practical experience in towing a delta-winged vehicle from a large aircraft, Project Eclipse was launched. The aircraft to be used as a stand-in for the Astroliner was a retired F-106 Delta Dart. Although much smaller than the projected aerospacecraft, the "Six" had a delta planform and was readily available, having been earmarked for use as a target drone under the QF-106 program. The NC-141A was likewise available, and was sufficiently powerful to take the former fighter aloft.

Prior to the tow missions, flights were made to gather data on the Starlifter's wake turbulence, which would be critical in understanding how the two aircraft would handle while tied together. These flights were carried out using both a NASA FA-18 chase aircraft and the QF-106 itself. The -106 in its guise as the EXD-01 received a ten line attachment mechanism forward of the cockpit;additionally, the nose mounted pitot probe was reduced in length.

Upon the completion of the Eclipse program, 61-2775 was finally retired and flown to Dover AFB to be put on display at the Air Mobility Command Musem.

NC-141A 61-2776 was the last of the quartet to be active, serving as the platform for the *Electric Starlifter* program, which was

Starlifter 66-0126 of the AETC, 16 October 1999. (Andy Thompson)

aimed at demonstrating "power by wire" technology that could ultimately replace purely hydraulic flight control actuation systems. Aircraft with such systems would benefit from not having large central hydraulic circuits that add much to weight, complexity, and maintenance requirements.

For the *Electric Starlifter* tests, -2776's ailerons were actuated by small, electrically actuated hydraulic systems, controlled by electrical signals generated by the control wheels of the pilot and copilot. The 418th began testing of -2776 on April 25, 1996, with the aircraft making an initial flight of an hour and a half from Edwards AFB.

After compiling more than a thousand successful flight test hours in the new configuration, -2776 closed at the program at Edwards on July 29, 1998. The *Electric Starlifter* was taken on charge at AMARC on August 7, 1998, ending more than thirty years of C-141A operations.

The NC-141As were not the only Starlifters to carry out test work. The 445th Airlift Wing of the USAF Reserve at Wright-Patterson AFB flew one of its C-141Bs fitted with a LIDAR (Light Detection And Ranging) system that could provide more accurate readings of winds to ground level by measuring the light reflected by layers of atmospheric aerosols.

(Andy Thompson)

The LIDAR Starlifter participated in a B-52 bombing test in Utah, passing along wind data for correcting aimpoints; aside from guiding bombers, an operational LIDAR system would also provide enhanced guidance for airdrop and gunship operations.

Chapter Two

The Galaxy Enters Service

Although the Starlifter would have a career stretching into the early 21st Century, the type did have one major drawback—its limited fuselage cross section, which was the same as the smaller C-130. This was one of the factors that kept the Starlifter from hauling all the items necessary for quickly deploying Army forces worldwide. Additionally, the C-141A's payload capacity was insufficient to haul such items as the fifty-ton M60 main battle tank, even had the fuselage been wider. A companion aircraft was needed, one that could use the same semi-prepared runways as the Starlifter, while carrying such loads as the M60, mobile bridging equipment, and other outsized and extremely heavy loads. The USAF projected a buy of around 150 aircraft, to compliment a roughly equal number of C-141As. As it turned out, C-5A production would amount to little more than half this figure, although 284 military Starlifters ended up being built.

The CX-4 requirement called for an aircraft that could carry 100,000 lbs over a 4,000-mile range. There was also the concept of a more advanced CX-X aircraft, one able to fly at least 10,000 to 12,000 miles without tanker support, and able to airdrop and air-retrieve heavy loads to cut reliance on forward airfields. Such abilities were well beyond the state of the art in the mid-1960s, and CX-X was primarily seen as a possible second-generation aircraft that would not be available until at least the late 1970s. Ultimately, the definitive CX-HLS was to haul a 100,000 lb payload some 5,000 miles, or 265,000 lbs for 2,500 miles. Additionally, the new design was to have a 30,000-hour design life, and the ability to operate from forward airfields with a minimum of ground support.

In contrast to today's military aircraft, which can take a decade or more to go from the drawing board (or CAD screen) to squadron service, putting the CX-HLS into service was given the highest possible priority. The need to replace C-124s and C-133s meant that the new transport would have to be operational by mid-1969. This goal would end up being missed, albeit not by much. However, the haste in getting the behemoth into squadron service would be one of the factors that nearly crippled the program.

Clearly the product of the same company, a C-141B of the 437th MAW and a C-5A of the 436 MAW share apron space at RAF Greenham Common on July 23, 1983. (Robbie Robinson)

The Galaxy's high-flotation landing gear was designed with idea of operating from substandard fields in mind, although testing showed that such operations caused damage to both the aircraft and matted runways. (Paul Hart)

C-5A 66-8307, the fifth Galaxy built, is seen here at RAF Mildenhall on March 15, 1986 while on the strength of the 436th MAW at Dover AFB, Delaware. (Robbie Robinson)

Another Dover bird, 68-0220, on April 14, 1985. (Robbie Robinson)

The idea of bringing huge transports into semi-prepared front-line airfields meant that much attention had to be paid to the landing gear design of the CX-HLS; from early on it was recognized that high-flotation gear with up to twenty wheels per side would be necessary. To test their design, Boeing called into use the venerable 367-80 *Dash Eighty* 707 prototype, fitting it with four additional wheels on each main gear, along with two extra on the nose unit. Douglas also conducted CX landing gear tests, in this case equipping a DC-7 with four wheels on each of its two main landing gear struts for taxi tests at Harper Dry Lake, California.

By the summer of 1964, Boeing, Douglas, and Lockheed were chosen as the CX-HLS study winners, and each was awarded a $400,000 contract to further refine their designs.

68-0212 flew with the 436th MAW in 1984. (Robbie Robinson)

C-5A 68-0225 of the Air Force Reserve Command's 439th AW formerly saw service with the 436th. It is seen here on March 27, 1998. (Robbie Robinson)

68-0212 after transfer to the ANG. (Paul Hart)

The next Galaxy off the line, -0213, equipped the 60 MAW of March 31, 1986. (Robbie Robinson)

L-500

Lockheed made maximum use of its C-130 and C-141A experience in designing the Galaxy, and at first glance the GL-500 was easily recognizable as a product of the Marietta airlift team. After having evaluated a number of designs with both straight and swept wings, the definitive Lockheed design was, at least outwardly, a "big C-141A," the high-mounted 25-degree swept wings with pylon-mounted engines and landing gear in fuselage pods were all features shared with the Starlifter. As on the L-300/C-141, the L-500 would have had a "T" tail, with Lockheed estimating that this feature would allow enough structural weight savings to permit over three tons of additional payload to be carried on some missions. The interior of the tail would be accessible via a laddered tunnel to ease maintenance.

The cavernous cargo hold was some 120 feet in length, 19.5 feet in width, and 13 feet in height at its highest point. This was slightly more capacious than earlier designs, which were just as high, but twenty feet shorter and three feet narrower. Making the fuselage wider allowed 463L cargo pallets or some vehicles to be carried side by side. 463L cargo handling features included centerline

C-5A 67-0170 of the New York ANG's 105th MAG, seen here in April 1987 after being repainted in European 1 colors. This aircraft formerly saw service with the 436th MAW (Robbie Robinson)

C-5A 69-0021 of the 137As/105 MAG (Paul Hart)

67-0173 of the New York ANG (Paul Hart)

rails for securing the pallets and floor rollers that could be turned over, providing a flat surface for other cargo. Projected loading time for three dozen pallets was an hour and a half.

The L-500 featured a visor nose design that allowed the whole nose to swing upwards on two points. This gave the advantage of easing ground loading, as well as avoiding a pressurized seam that would have to withstand the aircraft's velocity while airborne. With the visor up, the aircraft's engines could be run up, and loading was easier than with a clamshell configuration. A forward ramp, which made up part of the pressure bulkhead when raised, was some 22 feet in length when extended for loading. Straight-in loading from 54-inch high truck beds could be achieved; alternatively, the ramps could be lowered to allow drive-in access up an 11° incline.

The landing gear was as gargantuan as the rest of the aircraft, with each of the four main gear bogies having 6 wheels with a four-wheel nose gear, allowing operations in principal from forward airfields. A crosswind capability was provided, with a 20 degree right and left swivel. Uniquely, the gear was also "kneelable," allowing a parked Galaxy to lower or raise itself as necessary to ease loading of long pieces of cargo. Like the C-130 and C-141 before it, the Galaxy's main gear is housed in fuselage fairings, and these also contained air turbine motors and APUs to operate the cargo doors and kneeling mechanism, as well as to start the engines and provide heating and air conditioning while on the ground.

The flight deck and upper forward compartments are reachable via a ladder in the forward part of the hold; immediately behind the flight deck is a rest area with bunks for a relief crew, and aft of the bunks are twelve seats and a galley for couriers and other personnel.

Behind the wing structure and accessible by stairs from the cargo hold was the aft upper deck, with a galley and rearward-facing seats for a pair of loadmasters and 73 passengers.

Loads able to be carried by the Galaxy included:
A pair of M60A1 main battle tanks
Seven UH-1 "Huey" helicopters
Ten Pershing I tactical ballistic missiles and their launch vehicles
A pair of Minuteman ICBMs in their containers
Three partially dissembled CH-47 Chinook helicopters

Aircraft 69-0009 formerly saw service with the 60th MAW at Travis before being transferred to the ANG. (Paul Hart)

The C-5's very size makes hangarage difficult, although outside parking contributes to the deterioration of system. (Paul Hart)

"Blade inspection made easy." (Paul Hart)

The Galaxy remains the only USAF aircraft to have both a visor nose and rear door for roll-on/off cargo handling. (Paul Hart)

Although a fully "wet" wing would not be used, the C-5A's fuel capacity was nonetheless enormous, with a dozen wing tanks holding 332,500 lbs, or 51,450 gallons of fuel. Although the Galaxy would make little initial use of it, a refueling receptacle was fitted just aft of the flight deck to permit taking on fuel from KC-135As; the C-5A would be the first transport equipped from the start for midair refueling. Originally, the design included small "eyebrow" windows above the cockpit for additional visibility when approaching a tanker, but these were deleted before the Galaxy ever flew.

Although C-141s and chartered airliners would end up carrying most passenger missions, the Galaxy would have a secondary troop carrying capability, with 270 pallet-mounted seats able to be fitted in the cargo hold; plans for a dedicated troop carrier version with triple decks able to accommodate over 700 passengers never materialized.

An extensive avionics suite would be necessary for the aircraft to operate at low levels in all weather, fly in and out of airstrips with few or no navigational aids, and deliver precision airdrops. Housed in a radome at the tip of the visor nose, a multi-mode radar made of dual Ku- and X-band sets was fitted. Terrain avoidance and mapping of weather and ground features from high altitude would be the primary job of the X-band set, while the Ku-band unit would be tasked with approach and low-altitude mapping work, although in the case of one set malfunctioning, the other unit could take over at reduced performance.

A central new feature of the Galaxy's avionics fit would be the Lockheed-developed Malfunction Detection Analysis and Recording (MADAR) system, an onboard troubleshooting computer that would monitor hundreds of test points, detecting any abnormalities, alerting the flight engineer, and suggesting possible fixes. It was planned that MADAR would allow crews to radio ahead to have any needed Line Replaceable Units ready for installation; the C-5 was also designed to allow inflight maintenance of such items to be done in many cases. MADAR output came in both the form of hard copy printouts and a record on magnetic tape.

Engine Competition

While the competing CX-HLS airframe designs were to a large degree similar, differing mainly in details of weight and number of

105th MAG emblem, worn by 69-0012. (Paul Hart)

70-0453, seen on May 26, 1990. (Nick Challoner)

70-0467, at RAF Mildenhall May 1994. (Nick Challoner)

engines, the development of the powerplants themselves pushed the gas turbine industry to the cutting edge of technology, producing different concepts to meet the requirements.

At the time, the age of jet-powered airliners was less than a decade old, and low-bypass turbofan powerplants, exemplified by Pratt & Whitney's TF33 that powered the C-135B and C-141A, were state of the art. But for an aircraft the size and weight of the proposed CX-HLS, an entirely new engine would be necessary, one with up to 40,000 lbs of thrust, or better than twice the rating of the TF33. There were even designs that would not have made the step to turbofan power, using instead improved turboprops, such as the Allison T78 regenerative engine; these would have offered low-level endurance benefits.

What would eventually result in the TF39 engine for the C-5A was General Electric's tip turbine cruise fan engine effort. Although early planning for the CX-HLS had virtually ruled out the use of high-bypass engines, on account of their anticipated lengthy development time GE pressed ahead, and considered a number of arrangements for a production high-bypass cruise fan engine. One had a pair of gas generators mounted side by side atop a single fan. Another had a staggered arrangement of two generators, with a fan mounted on either side, while a similar design added an additional generator.

Actual CX-HLS engine hardware used technology from the company's GE-1 program, which had produced a number of prototype turbojet and turbofan engines. These included the GE-1/6, a 16,000 lb-thrust engine that was a subscale (two-thirds) prototype of the TF39, which won the CX-HLS engine competition in August 1965. The TF39's layout consists of a forward half-length 1.5-stage fan, through which most of the airflow passes, followed by a sixteen-stage compressor.

Pratt & Whitney also designed a series of demonstrator powerplants. The first of these was the STF-200C, a 31,000 lb thrust engine with a bypass ratio of 2 that was first run in the spring of 1964. The STF-200D derivative would have had a bypass of 3 and a thrust level of 34,000 lbs, while the -200F would have been rated at 39,200 lbs with a bypass ratio of four. A pair of STF-200Cs were built, and these were converted to JTF-14E configuration as proto-

On May 26, 1979 C-5A 69-0010 wore the emblems of both the 60th MAW and the bases' associate unit, the 349th MAW. (Robbie Robinson)

C-5A 68-0217 landing at Cairns, Australia in 1995. (Arno J.A.H. Cornelissen)

Their wing problems a thing of the past, the C-5A force continues to serve some three decades after the type's first flight, although their antiquitated avionics and older engines have lowered reliability rates. (Arno J.A.H. Cornelissen)

(Andy Spagna)

types for the C-X engine. The bypass ratio was increased substantially to 3.4:1, thanks to the substitution of a single fan stage for the -200's twin stage unit. The JTF14E was approximately 9.5 feet long and just over seven feet in diameter.

Despite losing the engine competition, Pratt & Whitney's program did not got to waste, as the company used the JTF-14E as the basis for the highly successful JT9D, which had a larger fan, and thus an increased bypass ratio of 5:1. JT9D variants would power many 747s and DC-10s, and have been candidates for C-5 reengining programs.

Flight trials of the TF39 were conducted using the ninth engine, designated XTF39, fitted to the starboard inner pylon of NB-52E 57-0119. Testing with the converted Stratofortress began on June 29, 1967. Early C-5 flights would use preproduction YTF39s, rated at 41,000 lbs thrust; these were later replaced by production models.

The Rohr Corp of San Diego was the subcontractor responsible for the Galaxy's engine pylons and nacelles, and was aided in

C-5A 68-0224 at Volk Field, 1997. (Tim Doherty)

(Andy Spagna)

(Andy Spagna)

68-0225 of the 337 AS/439 AW at Pease ANGB, 1997. (Tim Doherty)

C-5A 67-0167, June 1998. (Tim Doherty)

this effort by the receipt of a dummy TF39, shipped cross-country, fittingly enough in the cargo hold of a Hercules freighter.

The C-5 would be the only production aircraft to be powered by the TF39, but the engine would also form the basis for a commercial family of powerplants that would be in production well into the 21st Century. Early planning for a civilian version used the designation CTF-39, but by late 1965 the commercial model was known as the CF6. This was originally to be a close derivative of the military model, but evolved into a much more powerful engine.

The Galaxy is chosen

Secretary of Defense Robert McNamara announced on September 30, 1965, that Lockheed Georgia was the winner of the C-5A airframe competition, citing the company's "performance and price" superiority to the competing designs. At this time, the USAF planned to buy 120 aircraft, 96 of which would have equipped six squadrons, the others being used as attrition replacements and for training. Although the first five aircraft would carry out service trials, they would not carry the "YC" designation, as production tooling would be used in their construction and they would enter regular service after testing. The Galaxy would be the first new C-series aircraft to receive a designation in the new joint numbering system adopted by all the armed services in 1962; the C-1 through C-4 had been redesignations of Navy transports.

The C-5A was to break new ground in the area of program management; seeking to make the most of military budgets, and to provide an exact accounting of how much a weapons system had cost, McNamara had introduced the Total Procurement Package concept—a single price contract would be awarded to a prime contractor, which covered all costs of the program, from initial design to service introduction. This would prove disastrous, as TPP was poorly suited for a program that would push the edge of technology.

Access to the flight deck is provided by a ladder in the forward part of the cargo hold. (Tim Doherty)

Three decades after it rolled down the Lockheed assembly line, C-5A of the 337 AS was christened *City of Chicopee II*. (Tim Doherty)

Altus AFB received its first C-5A in 1969, and Galaxies are still based there as part of the 97 Air Mobility Wing of the Air Education and Training Command. 70-0453 is seen on May 28, 1995. (Tim Doherty)

One of the C-5's naval cargoes is the Mk.5 special operations boat, in this case belonging to a unit based at Cornado, California. The Galaxy is C-5A 69-0022. (Tim Doherty)

As had been the case on the Starlifter program, Lockheed had subcontracted out other major portions of the C-5 to firms all over the country. Subcontractors were present in all but six states, and even firms in the United Kingdom and Canada received Galaxy-related contracts.

Prior to the first flight, refinements to the basic design had to be made after wind tunnel tests showed the aircraft to have excessive drag. Fowler flaps replaced the original double-slotted units and the slats were enlarged. Smaller main gear bogies were substituted, allowing the pods to be made thinner; other efforts to cut drag included streamlining the nose and aft fuselage.

The Galaxy was unveiled to the public on March 2, 1968. President Lyndon Johnson was among the 40,000 visitors to the event, having flown in from his Texas ranch to take part in the occasion. Dwarfing a nearby Lockheed Jetstar and even Johnson's VC-137

The two mainstays of Lockheed's transport stable. The Galaxy is 69-0020 of the 439 AW; the Herks are C-130Hs of the Georgia Air National Guard. (Tim Doherty)

The Galaxy's T-tail is large enough to accomodate an internal ladder to allow maintenance. (Tim Doherty)

C-5A 67-0173 of the NY Air Guard conducting touch and go's at Wheeler-Sack Army Air Field, Fort Drum, New York on September 18, 1998. (Tim Doherty)

69-0013 prepares for loading at Otis ANGB, August 22, 1999. (Tim Doherty)

Aft side doors allow for personel loading via airstairs; although not primarily a troop carrier, the Galaxy can be fitted with 270 seats in the cargo hold, in addition to seating in the upper compartments. (Tim Doherty)

"Air Force One," the C-5A was displayed alongside military vehicles and other representative cargoes.

June 26 was the date of the first taxi tests; Leo Sullivan, still Lockheed's chief test pilot, headed up a five man crew. However, a loss of pressure in the number three hydraulic system forced a premature end to the test. Sullivan was again at the controls four days later when the first flight was made, lifting off from Dobbins at 7:47 AM for a 94-minute flight. The Galaxy was fitted with a large nose-mounted instrumentation boom, and bore the nonstandard script lettering "Galaxy" on the forward fuselage.

On August 16, 1970, a C-5A set out from Edwards on a flight intended to simulate a long-range ferry mission. Operating with a full load of fuel, the unladen Galaxy stayed in the air for over twenty hours without tanker support, flying first to Maine, then to Georgia, and then finally back to Edwards.

One aircraft conducted tests to determine the Galaxy's inflight icing characteristics, flying below and behind an NKC-135A as the modified Stratotanker sprayed water from its refueling boom. Aircraft -0168 flew from Pope AFB with over 198,000 lbs of armored vehicles as its payload, consisting of an M577, a 175 mm self-propelled gun, and an M60 MBT. And, by the spring of 1970, one aircraft was flying with a slat made of boron composites. Compared to aluminum slats, the boron unit was 50 pounds lighter and of simpler construction.

Also investigated during the Galaxy test program was the effect of the giant transport's wake turbulence on smaller aircraft. Testing over Edwards was conducted by flying U-3 and F-104 aircraft into a C-5A's wake; using the data gathered the FAA in January 1970 issued interim guidelines for aircraft operating around C-5s and 747s.

Entry Into Service

When the C-5A program had begun, the Military Air Transport Service had been the intended user command, but this was changed when MATS was replaced by the Military Airlift Command on January 1, 1966. The command's training unit, the 443rd MAW, would be the first to receive the C-5A, the unit having been moved from its previous home at Tinker AFB to Altus AFB, Oklahoma, a former SAC base. The first Galaxy for the 443rd, *Spirit of Altus,* arrived on December 17, 1969.

The 436th MAW at Charleston AFB, South Carolina, would be the first operational unit, with MAC commander General Jack Catton piloting the first aircraft to the base on June 6, 1970. The occasion was marred, however, by a minor but very public mishap. Upon landing, one of the left main gear wheels came off, damaging another and then rolling down the runway after the aircraft. The Galaxy came to a stop with little damage, and in other times little

(Chris Reed)

C-5A 70-0446 of the 60th MAW, July 5, 1982. (Masanori Ogawa)

might have been made of the incident, but the media of the day made much play of the errant wheel.

The 60th MAW at Travis AFB, California, was to be the West Coast C-5A unit; the 75th MAS, a former C-141A operator, was the wing's first Galaxy squadron, and was joined in 1972 by the 22nd MAS. C-5 deliveries were completed in May 1973 with the hand-over of the 81st aircraft.

Quite aside from the wing problems detailed elsewhere, the Galaxy, like any other new aircraft, encountered its share of early technical difficulties. The pneumatic kneeling system, which had proven liable to overheating, was replaced by a slower but more reliable hydraulic system. The gear electrical systems were also switched to a less troublesome solid-state configuration.

May 1970 was a bad month for the Galaxy program, as there were two serious accidents in as many days. On May 24, an aircraft of the 443rd was taking off from Marietta bound for Oklahoma when electrical problems forced a return. Landing at night with no lights, the crew managed to get the aircraft on the ground, but one of the main bogies did not fully cycle, and was damaged on touchdown. The nose gear also failed during the landing.

Even more seriously, the following day saw the loss of a Galaxy. Aircraft 67-0172 (the 11th airframe) was taking off from Palmdale, California, when a nosewheel steering problem forced an abort, and a fire broke out. The crew managed to evacuate after the C-5A came to a stop, but several firefighters were hurt fighting the blaze.

Another Galaxy was lost on October 17, 1970. After having developed a fuel leak during a flight the previous day, the first aircraft was undergoing a fuel cell purge in the early morning hours on the flight line at Marietta when an explosion occurred. Two others followed, tearing the aircraft to pieces, and started a major fire, killing a flight line technician.

By 1970, the USAF had changed the C-5A's operational concept; gone were notions of flying the aircraft at low level into the forward areas of battle zones. Galaxies would have to use long, hard-surface runways, although operations from smaller fields were possible in emergencies. Testing had shown that the aircraft's radar was inadequate for low-level use, while rough field testing did damage to both runway matting and the aircraft's own hydraulic systems. Gone also was the idea of using the Galaxy in the troop carrier role, as these missions could be flown by chartered airliners, freeing up the C-5As to carry outsized equipment that no other aircraft could.

Cost Troubles

The C-5A program was to suffer cost overruns that paralleled the aircraft's physical size. The original contract was valued at around $1.9 billion, but by then there were several reasons for the drastic increase in costs. The 1967 redesigns had all incurred additional expenses; had a heavier aircraft been permitted, costs may have been lesser. Another factor was the era that the Galaxy took shape in; with the conflict in Southeast Asia raging and the American space program in full swing, aerospace designers, labor, and materials were all in short supply, and Lockheed had to pay top dollar for each.

In late November 1968, only two months before the scheduled decision on the 57 Batch 2 aircraft, Senator William Proxmire of Wisconsin had the General Accounting Office begin an investigation of the C-5A contract. By this time, the cost of buying the initial 58 Galaxies was estimated to have exceeded $3 billion, well over a billion more than foreseen at the start.

In January 1969 the USAF indicated that the Batch 2 buy would number only 23 aircraft; many critics of the program had called for the second run to be canceled altogether. The reduced order was confirmed in November 1969.

The late 1960s and early 1970s were extremely bad times for Lockheed, a situation only partially caused by the C-5A's troubles. The L-1011 Tristar, being built by the company's California division, ran into severe problems when engine supplier Rolls-Royce went into bankruptcy, creating a drain on Lockheed's coffers. Additionally, the AH-56A Cheyenne attack helicopter was in trouble, as were several other programs. Ultimately, and in a storm of controversy, the U.S. Government guaranteed several hundred million dollars in loans to Lockheed from commercial banks to help keep the company from going under.

Commercial Models

Despite the increasing specialization in the design of military transports, at the time of the CX-HLS program's inception there were serious plans of using the resulting aircraft as the basis for a commercial variant. While later efforts would concentrate on selling such aircraft as commercial freighters, at the time of the program's conception there was interest in using the resulting aircraft as an ultra-large capacity airliner, carrying up to 1,000 passengers, which could radically cut the cost of air travel.

Early L-500 configurations included the L-500-1 version, which was dedicated to passengers. The rear ramp and visor nose would be deleted, and with no need to land on austere fields, lighter twin tandem six-wheel main gear would be fitted. Up to 900 passengers could be carried in a high-density layout, while a more luxurious 667-capacity configuration was also possible. The L-500-2 combi version, which reverted to the C-5A's visor and ramp configuration, would carry both passengers and freight; the upper deck would accommodate 225 passengers, while the lower two would carry up to 219,000 lbs of cargo. And finally, the L-500-3 would have dispensed with all passenger capability, raising its cargo payload to 242,000 lbs.

Later L-500 plans were predominantly for all-cargo versions, as by the late 1960s all three major aircraft manufacturers were concentrating on meeting the large airliner market with purpose-designed widebody designs; Lockheed's entry being the L1011 Tristar. Ironically, this would be the only of the three that was not powered (in at least some models) by the CF6 derivative of the TF39, instead using British Rolls Royce RB.211s.

The L-500-107C concept of 1966, powered by advanced 50,000 lb-thrust powerplants, would have carried 330,000 lbs of cargo in a triple-decked configuration.

By the next year, Lockheed had four commercial Galaxy models defined. None of these would be powered by the TF39 or CF6, relying instead on either the Pratt & Whitney JT9D-7 rated at 45,000 lb thrust, or an uprated version delivering 47,500 lbs. There would be two standard length versions, the L-500-114M and -114P. The M would gross out at 818,500 lbs (282,400 lbs of which would be cargo). The P would use the more powerful engines, with a payload capacity of 280,500 lbs. In paralleling studies on stretching the C-130/L-100 and the C-141A to allow these aircraft to carry more volume-intensive cargo, Lockheed also proposed a pair of Galaxy models, the L-500-114Q and -114R, which would have been lengthened by ten feet. Payload would have jumped to at least 357,000 lbs.

Although the later L-500s would have used the basic Galaxy planform, there would be major structural differences when compared to the military model. There would be some deletions of unneeded military features; with no need for roll-in/roll-off loading, the aft cargo doors would not be fitted, thus freeing up over twenty additional feet of space for cargo. The kneeling landing gear would also be dispensed with.

In order to make maximum use of the available volume, smaller containers would be carried suspended from the cargo hold ceiling, and the upper compartment would be some eight inches higher than that of the C-5A, with a pair of portside doors for the loading of cargo containers.

Lockheed saw the L-500 as the largest part of an interlocking air transport system, with trucks bringing cargo to feeder airports, where L-100s would pick it up for transhipment to major hubs where L-500s would be waiting, the whole process going into reverse at the other end of the journey. As no contemporary airport facilities could handle aircraft the size of the Galaxy, Lockheed devised a new terminal design, resembling a "T" in overhead planform. Loading and unloading of cargo from trucks would take place at docks in the "head" of the structure.

Lockheed was not the only company designing infrastructure for the proposed new freighter. Avco Aerostructures, a major C-5A subcontractor, proposed a new means of rapidly converting cargo aircraft to passenger configuration. This would be achieved by the AvLounge, a self-propelled modular cabin that could be driven on and off aircraft as needed. Three 100-seat AvLounges could be carried by a Galaxy, but smaller Lockheed aircraft would also be compatible with the system, the Starlifter carrying two, and the stretched L-100-20 and -30 Hercules able to accommodate one.

In the early 1970s Universal Airlines proposed to buy five commercial Galaxies as the centerpiece of a plan to offer transcontinental automobile/passenger flights. United and TWA had also expressed interest in using the L-500 as a car carrier, principally for moving new luxury cars from assembly lines in Michigan to California much faster than would be possible by rail. Lockheed estimated, that in a triple-deck configuration, a car carrier Galaxy could haul over 100 of what at the time were termed "economy import" automobiles.

Nothing came of such plans, and afterwards little was heard of the commercial Galaxy. The extensive changes necessary from the military version would have meant that redesign costs would have had to be passed on to any customers, while competing freighter versions of the 747 were available from an active production line.

In 1975, a C-5A did take part in the ground tests of *Project Intact* (Intermodal Air Cargo Test), an effort to establish the design parameters for a next-generation airlifter optimized for carrying intermodal cargo and over-the-road trailers. By this time the Galaxy was out of production, and was only being used as a representative aircraft.

C-5A 70-0457, 60th MAW, August 29, 1983. (Masanori Ogawa)

Operation *Nickel Grass*

The C-5's first real operational test came during October 1973, when Israel found itself besieged by Arab armies attacking from all directions. Despite warning signs, the Israelis had been taken by tactical surprise by the offensive, rapidly losing ground on the Sinai Peninsula and elsewhere. Losses in men and material were heavy, and stocks of ammunition were rapidly drawn down. With Israel's conventional military forces nearing collapse, the government was faced with the impossible choice of either surrendering or unleashing the country's unacknowledged but very real stockpile of nuclear weapons.

While trying to keep the U.S. out of direct involvement in the war, President Nixon nonetheless ordered MAC to carry out an emergency airlift, codenamed *Nickel Grass*, in the hopes that with a renewed supply of war material the Israelis could themselves turn the tide of the conflict.

Hoping to cut off Israel from easy reinforcement from the U.S., Arab states had warned of an oil boycott of any nation overtly or otherwise aiding their adversary. Most NATO members, fearing a cutoff of their fuel supplies, denied the U.S. permission to use their fields for the operation. However, Portugal turned a blind eye towards transports staging out of Lajes Field in the Azores.

Lajes was absolutely critical to the operation; Galaxies could still have reached Israel from the U.S. nonstop, and without tanker support, but the payload would have been drastically shrunk to twenty tens per flight. Despite being equipped with refueling receivers, the Galaxy force was at the time mostly unable to use this capability, as fewer than two dozen C-5A pilots were qualified to take on fuel from tankers.

Upon entering the Mediterranean, U.S. airlifter crews were essentially flying over a powder keg; below, U.S. and Soviet naval forces were jockeying for position, while North African Arab countries could potentially send MiGs after the defenseless transports. Carrier groups operating in the Med provided the Galaxies and Starlifters with protection, while other vessels stood by to provide SAR if necessary. While having to stay overwater, the transports also had to take a more northerly course than desired to stay in southern European FIRs, and out of those of Casablanca, Tunisia, Tripoli, and Cairo.

Although no attacks were carried out on American aircraft, Arab fighters were sighted on occasion, and tensions were running high. Upon nearing Israel, the C-5s were met by IAF Mirage III and F-4 fighters for the final run in. Even once on the ground at Tel Aviv's Lod Airport the transports were not completely out of danger, as fighting was taking place only a hundred miles from the field, and the possibility of Arab airstrikes, commando operations, or *Scud* missile attacks could not be discounted.

Israeli aircraft, which had performed so admirably in June 1967 and in skirmishes thereafter, were now falling in large numbers to the improved Arab air defenses, which included top of the line Soviet SA-6 SAMs and ZSU-23-4 AAA tanks. Israeli losses were hurriedly replaced by the delivery of A-4s and F-4s directly from U.S. squadrons and reserve stocks, and the airlift brought over new ECM equipment to deal with the threat. Although deliveries of fixed-wing aircraft to Israel entailed self-deployment or shipment by sea, C-5s did transport A-4 rear fuselage sections, as the Israeli Air Force had a number of Skyhawks that had their tail areas shredded by hits from SA-7 shoulder-fired SAMs. Galaxies also airlifted CH-53 heavy lift helicopters to enhance Israeli airmobility. Israeli tank losses were also high, and the Galaxies delivered around 30 M60A1s to Tel Aviv. The delivery of new types of weapons were also necessary; within days, the U.S. had delivered at least 2,000 BGM-71 TOW anti-tank missiles; also sent over were AGM-65 Mavericks to arm Israeli Phantoms and Shrike anti-radar missiles to combat Arab SAMs.

Although the main aerial ports in the U.S. were Charleston, McGuire, and Dover, USAF transports were active throughout the country, visiting stockpiles and even weapons production lines to pick up war material.

The long duration of the missions vindicated Lockheed's provisions for relief crews during the Galaxy's design. A typical mission carried an extra pilot, navigator, and flight engineers; additional loadmasters were also carried to keep the critical job of quickly unloading at Tel Aviv as smooth as possible. From early on in the operation, an Airlift Control Element (ALCE) was in place at Lod to coordinate USAF operations, and to provide aircraft with needed maintenance.

While Ni*ckel Grass* was underway, the Soviets were also attempting to reinforce their Arab allies, by both air and sealift. At the time, there was no operational jet airlifter available to the VT-A, although it could call upon the Antonov An-22 *Cock*, an outsized turboprop transport, that had, until the C-5A, been the world's largest military aircraft. Together with smaller An-12 *Cubs*, the An-22s operated from Kiev to Damascus Syria. Ironically, both U.S. and Soviet aircraft were transiting Yugoslavian airspace, and the Antonovs also used the airspace of Turkey, a NATO ally.

67-0169 of the 60th, seen on November 14, 1983. (Masanori Ogawa)

66-8307 of the 436th MAW, March 8, 1984. (Masanori Ogawa)

Although the Soviet airlift had the advantage of being much shorter in length than the U.S. operation—less than two thousand miles—it also accomplished less, moving approximately 15,000 tons, despite flying nearly a thousand missions.

Equipped with fresh supplies and new weapons, Israeli forces rallied, crossing the Suez and penetrating deep into Egypt, cutting off Sadat's forces. Coupled with successes on the Syrian front, the war had turned in Israel's favor, and soon the Arab capitols of Damascus and Cairo were themselves under threat. The dramatic reversal in battlefield fortunes led to Soviet preparations to overtly intervene themselves, with Red Army airborne forces being put on alert, and some units being forward deployed to Yugoslavia.

Seeking to head off the specter of a Soviet-Israeli clash that would inevitably draw the U.S. in, as well, President Nixon, on October 24, ordered selected forces to raise their combat readiness to DEFCON III. This had the desired effect of showing U.S. resolve against Russian involvement, and a lasting, albeit tense cease-fire went into effect on October 28. Although SAC and air defense forces had lessened their alert states not long after they had been raised, forces in Europe remained on alert until the end of October, and carrier groups in the Mediterranean did not return to peacetime levels of readiness until well into November.

Despite the cease-fire, tensions remained high, and *Nickel Grass* continued for several weeks, finally terminating on November 14. C-5As had delivered 10,800 tons of cargo in the course of 145 missions, with Starlifters flying another 422 sorties. Altogether, 22,395 tons had been carried by the airlift.

Export Hopes

Although the USAF has been the sole operator of the C-5, export sales of the Galaxy have been considered several times. One of the first prospects for a sale came in late 1974, when Lockheed and Iran announced tentative plans to supply the Iranian Air Force with ten Galaxies from a restarted line. Costs of resuming production were to have been borne by the Iranians, thus opening the possibility of the USAF being able to buy additional aircraft from a "warm" production line. The sale never went through, and the Iranians ended up buying a tanker-transport version of the 747.

In 1970 the RAF had considered the purchase of a small number of C-5As; although by that time Britain was well on its way from withdrawing from most non-NATO overseas obligations, it was thought that a new government might reverse this course, leading to a need for heavy airlifters for missions to the Far and Middle East. To make the Galaxy more attractive to the British, Lockheed proposed a version with Rolls-Royce RB211 engines, a concept that would resurface more than two decades later.

Vietnam

The Galaxy entered service after the height of U.S. involvement in Vietnam, but would nonetheless carry out some notable missions to Southeast Asia. In April 1971, a Galaxy took three CH-47s from Pennsylvania to South Vietnam.

Routine C-5 operations into the theater began in the spring of 1972, following the Easter Invasion of South Vietnam by conventional NVA forces. The widespread communist use of armor, including T-54 tanks, led to a hurried effort to get countering anti-tank assets into South Vietnam, including TOW-armed Huey Cobra gunships flown in on C-5s. Tanks were also brought into Da Nang to take part in fighting near the base itself.

The ending of U.S. combat involvement in Vietnam in early 1973 led to *Operation Enhance Plus*, an effort to rapidly strengthen the South Vietnamese armed forces as quickly as possible before the arms transfer limitations of the Paris peace deal went into effect. *Enhance Plus* included the airlift in of F-5s and helicopters.

The Galaxy's most well-known Vietnam mission was also one of the most tragic stories of the war. In early April 1975, the situation in South Vietnam was increasingly desperate, as a new NVA offensive was rapidly advancing towards Saigon. This time there was no real hope of American military power coming to the rescue, and the U.S. was preparing to evacuate its citizens. Well before this operation, *Frequent Wind*, was launched, *Operation Babylift* was put into place to get Vietnamese orphans out of the country and to adoptive parents in the U.S.

The first *Babylift* mission was launched on April 4, using a Galaxy to take over 150 children to safety. None could foresee that this mercy flight would end in disaster almost as soon as it began.

Captain Dennis Traynor and his crew, flying C-5A 68-0218 of the 60th MAW, left Tan Son Nhut AB in Saigon, bound for Clark AB, the Philippines. For a short while the flight was uneventful; this changed dramatically when an explosion shook the aircraft, ripping the aft doors and ramp off. Debris tore the rudder and elevator control cables, and two of the hydraulic lines were also cut. Using engine power and one aileron, the crew struggled mightily to turn the aircraft back towards Vietnam, while attendants, dazed and injured themselves, fought the effects of the decompression to aid the children.

Traynor's crew managed to get back over land, but their luck ran out before they could make Tan Son Nhut, when the aircraft had to be force-landed in a rice paddy a few miles from the base.

It was speculated at the time that the C-5A might have been the victim of a shoulder-fired SAM, but the cause of the disaster was attributed to the failure of a door lock, which in turn caused the other locks to give way as well. Despite the tragedy, the effort to evacuate orphans continued, with several thousand reaching safety before Saigon finally fell.

Disaster Relief

Although the Berlin Airlift remains the best known of the USAF's humanitarian operations, over the decades MAC, and later AMC, have carried out many missions of mercy to all corners of the globe. Since the early 1970s, the C-5 fleet has participated in many of these, carrying everything from food and blankets to firetrucks.

On February 4, 1976, a massive earthquake hit Guatemala, causing major damage, and by the next day U.S. military and civilian aircraft were bringing relief supplies into the stricken country. A pair of C-5As took part, bringing in communications equipment and purification equipment to provide clean drinking water.

Despite its tremendous size and weight, on occasion the C-5 has itself been proven vulnerable to natural disasters. On May 11, 1982, Altus AFB, OK, was struck by a tornado, with the storm blowing the nose of one 443rd MAW Galaxy into the starboard wing of another. Winds also damaged three other C-5As and five C-141s.

Not all disasters that Galaxies have responded to have been natural. As part of the federal government's response to the March 1979 nuclear accident at the Three-Mile Island plant near Harrisburg, Pennsylvania, a Galaxy brought in material to help contain and clean up radioactive material from the partial meltdown. And almost ten years to the day after TMI, C-5s were part of the MAC airlift bringing in cleanup supplies to Alaska following the disastrous Exxon *Valdez* oil spill at Prince William Sound.

With the eastern coast of the U.S. under threat from Hurricane *Floyd* in September 1999, the USAF staged a massive movement of aircraft from coastal bases to points further inland. Robins AFB, Georgia, was one of the affected bases, sending a pair of Galaxies to Grissom ARB, Indiana. A C-5 also helped in evacuating other aircraft; in the hurried effort to clear Patrick AFB, Florida, near the projected path of the storm, the 920th Rescue Group shipped out a pair of its HH-60s aboard an AFRC C-5A that had been diverted to the base.

In August 1999, an earthquake hit Turkey that severely damaged much of the country's infrastructure, killed thousands outright, and trapped many others in ruined buildings. Even as major aftershocks rattled the region, the USAF began sending relief flights, including a 436 AW Galaxy carrying SAR teams and other materials. In order to get much-needed relief to Turkey as quickly as possible, the flight was made direct, with a pair of KC-10 Extenders "topping off" the C-5 in flight.

Mere weeks after the Turkish earthquake, a 7.6 tremor hit near Taipei, Taiwan, on September 21, killing more than 2,000 people. A Galaxy left Dover the following day, carrying 30 tons of equipment, personnel, and vehicles from the Fairfax County SAR team.

Other humanitarian missions by C-5s have included:

Mexico City Earthquake: 1985.

Operation Fiery Vigil: Evacuation of Clark AB and the Subic Bay naval base in the Philippines following the eruption of Mt. Pinatubo, June 1991.

Operation Provide Hope: Transport of relief supplies to states in the former Soviet Union.

Hurricane Andrew: August 1992.

Hurricane Marilyn: Relief to the Caribbean, September 1995.

Typhoon Paka: Guam, December 1997 to January 1998.

Hurricane Georges: September 1998.

Hurricane Mitch: Relief to Central America, November 1998-January 1999.

Venezuelan Mudslides: December 1999.

Airlifting Aircraft

Hawker-Siddely XV-6A: The very first aircraft to be carried by a Galaxy was an XV-6A Kestrel, which was brought from Edwards AFB to the USAF Museum at Wright-Patterson AFB in 1970.

Boeing B-17 Flying Fortress: C-5s have been used to airlift Flying Fortresses on several occasions. When the combat veteran B-17G *Shoo Shoo Baby* was returned to U.S. control in the early 1970s, the Fort was taken apart in preparation for shipment to the USAF Museum; the airlift was carried out by a C-5A on June 14, 1972. Subsequently restored at Dover AFB, "Baby" was flown back to Wright-Patterson, while Dayton's former display Fort, a non-airworthy DB-17, was broken down and shipped to Dover aboard a Galaxy.

69-0011 of the 60th MAW. (Masanori Ogawa)

Boeing KC-97 Stratofreighter: On October 20, 1999, a C-5 delivered the nose section of KC-97 51-230 from Beale AFB, California, to Dover AFB, Delaware, where it will be reassembled for display at the Air Mobility Command Museum.

Consolidated B-24 Liberator: In 1999, a Galaxy airlifted a disassembled B-24 to RAF Mildenhall for restoration and display at the American Air Museum.

Convair F-102 Delta Dagger: The McClellan AFB Museum's "Deuce" was airlifted to the base in 1984 by a C-5A.

Fairchild-Republic A-10 Warthog: Although the prototype YA-10 had been taken to Edwards by C-124s, the Galaxy also provided airlift support to the program, with a C-5A transporting the initial Development Test & Evaluation aircraft to California. Whereas the prototype Warthog's fuselage had to have its engines removed to fit inside a Globemaster, the Galaxy was large enough for these to remain installed.

General Dynamics F-16: The prototype YF-16, 70-1567, was loaded onto a C-5A on January 8, 1974, for transport from the General Dynamics (formerly Convair) plant at Fort Worth, Texas, to Edwards AFB, California, for its first flight.

Grumman HU-16 Albatross: Moving Albatross 51-7209 from Luke AFB to the museum at McClellan for display was accomplished by a Galaxy flight in March 1998.

Lockheed F-117: Another product of the legendary "Skunk Works," the F-117 was even more secretive than its U-2 and SR-71 predecessors, and MAC airlifters were used to carry completed aircraft from Palmdale to Groom Lake, and later Tonopah, Nevada, under tight security. This operation was preceded by the C-5A airlift of the two *Have Blue* demonstrator aircraft.

Lockheed Martin F-22 Raptor: Galaxy support of the Raptor program has included flights to move one YF-22 to Rome Labs for electromagnetic testing, the other prototype to Wright-Patterson AFB for display at the USAF Museum, and the first F-22A to Edwards AFB, this last mission having taken place in February 1998. Subsequent Raptors have self-deployed to California.

Lockheed Martin P-3 Orion: Following a crash of an EP-3E at Souda Bay, Crete, in September 1997, a Galaxy transported the ELINT aircraft's fuselage back to the U.S.

Lockheed T-33 Shooting Star: Another Lockheed product, albeit one from the California division, a T-33 was lifted by a C-5 from Lackland AFB, Texas, to the Air Mobility Command Museum at Dover, Delaware.

Lockheed A-12 Blackbird: The first retirement of the USAF's SR-71s in 1990 allowed some of these exotic aircraft, and their A-12 predecessors, to be released to museums. Some of these were moved by Galaxies, including A-12 60-6931, which was taken by the New York ANG from Palmdale, California, to the Minnesota Air National Guard Museum.

McDonnell Douglas F-15 Eagle: The prototype F-15, 71-0280, was transported by a C-5A to Edwards from St. Louis prior to the start of flight testing in July 1972. The next four Eagles were also brought to California by Galaxies.

Civilian Aircraft: Not all aircraft carried by Galaxies have been military: in late July 2000 the U.S. Acrobatic Team was transported to Europe aboard a C-5 of the 433rd Airlift Wing.

Boeing Vertol CH-47 Chinook: Coming into service in the early 1960s, the HC-1B/CH-47 Chinook was the U.S. Army's first turbine-powered heavy lift helicopter, and was a crucial component of the new strategy of airmobility. As such, it was from the start a prime cargo for transport by the C-5, and CH-47As were used during early fit tests of the competing mock-ups. Overseas deployments of Chinooks have often occurred aboard Galaxies, as most CH-47s are not equipped with aerial refueling capability; deliveries of export models, such as those going to China, have also been conducted using C-5s.

Fairchild Republic T-46: A C-5A brought the prototype T-46 New Generation Trainer from the Republic assembly plant at Farmingdale, New York, to Edwards AFB on August 29, 1985.

69-0001 of the 60th, in "lizard" colors on October 23, 1988. (Masanori Ogawa)

Mikoyan-Gurevich MiG-25 Foxbat: Among the more unusual cargoes ever hauled by a Galaxy was the MiG-25 *Foxbat* interceptor that was flown by Soviet Air Force Lt. Victor Belenko on a defection flight to Hakodate, Japan. With tensions running high over Western acquisition of the advanced fighter, the *Foxbat* was partially dismantled, with the fuselage being moved by a 60th MAW Galaxy to the Hyakuri Air Base near Tokyo. The enraged Soviets, attempting to cow the Japanese into quickly returning the MiG, launched aircraft that came near to Japanese airspace; to guard against any attempt to intercept the C-5A, the JASDF escorted the transport with more than a dozen F-104J and F-4EJ fighters. The Galaxy landed without incident, as did a JASDF C-1 transport carrying the MiG's other components.

Northrop Grumman B-2 Spirit: Another highly secretive aircraft program that the Galaxy supported was the B-2 "stealth bomber." Northrop had built a large facility at Palmdale, California, for final assembly of the B-2, but major subcontracting work was done elsewhere and the components shipped in. For example, in August 1987, a C-5A brought a wing set to Palmdale.

Northrop F-5E Tiger II: As Northrop's improved F-5E export fighter entered series production in the mid-1970s, the USAF used C-5As to deliver the aircraft to customers worldwide. As the F-5Es were not equipped with provisions for aerial refueling, delivery flights would have entailed numerous stops. With their wings and horizontal stabilizers removed, up to eight Tiger IIs could be carried side-by-side inside a Galaxy.

Rockwell/MBB X-31A: Following its exhibition at the 1995 Paris Air Show, one of the U.S./German Enhanced Fighter Maneuverability demonstrators was returned to Edwards AFB by an Air Force Reserve Galaxy.

C-5A 70-0457, 60th AW, August 22, 1993. (Masanori Ogawa)

69-0007, 433rd AW, April 27, 1997. (Masanori Ogawa)

Wing Problems

While the financial difficulties incurred by Lockheed over the Galaxy program nearly doomed the company, another major problem would soon arise, also threatening to cut short the C-5A's career. As part of the post-1965 effort to lighten the design sufficiently to meet weight requirements, much material was removed from the wing structure, compromising its integrity and dashing any hopes of meeting the aircraft's 30,000-hour life requirement.

The C-5A's wing problems began on July 13, 1969, when a wing crack occurred during testing of the Galaxy test airframe at Marietta. The structure was supposed to have endured up to 150% of its design stress limit, but failed when a crack developed near the outboard pylon at only 128% of load. Several months later, another crack occurred at an even lower load, but it was thought that relatively minor reinforcement would fix the problem.

However, further wing flaws were to surface in January 1970, when an eight to ten-inch crack was found on a Galaxy at Marietta, leading to the fleet being grounded for two days while inspections were carried out. The aircraft in question had flown 84 missions, accumulating just under 300 hours in the air. Ultimately, it was found that the C-5A force had an effective lifetime of around 8,000 hours, only a fraction of that required, and low enough that the USAF would have to begin retiring the type relatively soon.

Pending a definitive fix for the problem, several measures were taken to lessen wing fatigue. Firstly, a payload restriction of 50,000 lbs was enacted, although this was a peacetime limit and could be waived in times of war or other emergencies. Secondly, the fuel system was re-rigged so that the outboard tanks were drained last, thus keeping wing flexing to a minimum. The fitting of an active aileron system also dampened wing movement.

These were only stopgap measures, however, and throughout the early 1970s Lockheed studied fitting the C-5A fleet with either wing boxes, or entirely new wing structures.

Finally, the beginning of what would be the *Pacer Wing* retrofit started in December 1975, with Lockheed being contracted to

70-0453 at Yokota, May 18, 1999. (Masanori Ogawa)

design new wing boxes. Avco began construction of a pair of the new-design boxes in 1978; one set was refitted to C-5A 68-0124, which began testing in the summer of 1980, while the other was used for ground tests. These were successful, leading to the program for Lockheed to fit the new wing boxes to 77 C-5As. Early plans had called for the existing center section boxes to be kept, but these ended up being replaced, as well.

Even the rewinging was politically controversial, as critics claimed that for the estimated $1.34 billion that refitting the entire fleet would cost, it would be possible to buy new DC-10s, L-1011s, or 747s. Ultimately, the USAF decided that rewinging existing Galaxies would be the most cost-effective solution to the airlift problem.

Fitting the new wings entailed each aircraft spending approximately 120 days at Lockheed. The first step in the process entailed removing the engines and their pylons, the control surfaces, panels, and draining and cleaning the tanks. The second step involved the removal of the inner and outer wing sections, and finally the center section. Installation of new structures began in reverse order, with the new center wings being fitted, followed by the inner and outer wings, and the engines, pylons, fairings, and panels, and a final check-out.

C-5A 67-0173 was the first example to undergo a "production" rewinging, which was completed in February 1983. In a demonstration of the type's renewed capabilities, a rewinged C-5A broke world weight-carrying records on December 17, 1984. Taking off from its birthplace at Marietta, the Galaxy took a 232,477 lb payload to above 2,000 meters. Additionally, at an overall weight of 922,000 lbs, this particular C-5A was the heaviest non-LTA aircraft to ever fly up to that point. *Pacer Wing* finally wound down in July 1986 with the delivery of the last rebuilt aircraft to Dover.

The wings were not the only C-5 structures to encounter problems early in the program. An accident in the fall of 1971 at Altus saw a TF39 engine fall off its pylon; fortunately, this occurred while the aircraft in question was on the ground, having its engines run up in preparation for take-off. High flight time Galaxies were grounded for inspection, but this was extended to the rest of the fleet when a pylon crack was found on a low-hour aircraft.

C-5 Detail Section

View from just behind the forward ramp, just below the flight deck. (Chris Reed)

Side view of the flight deck, with the visor nose raised. (Chris Reed)

Bottles that are part of the cargo compartment fire suppression system are located at the top of the compartment walls. This unit, near the aft doors, has a light next to it. (Chris Reed)

Interior view of the aft door area. (Chris Reed)

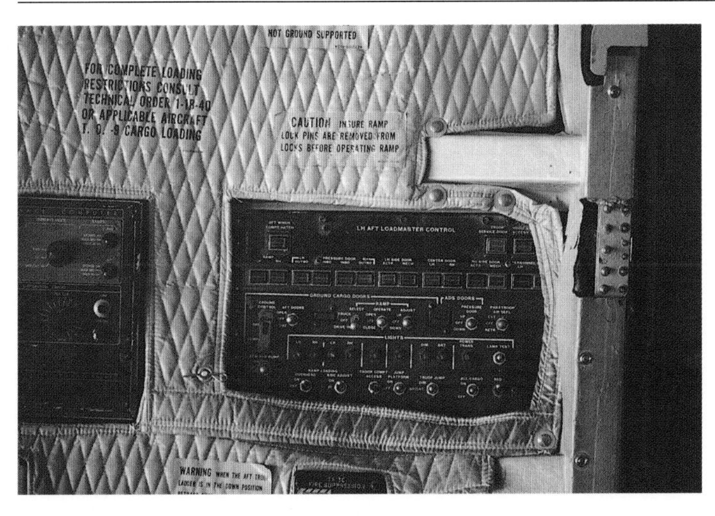

Aft loadmaster control panel and balance computer, near the rear ramp. (Chris Reed)

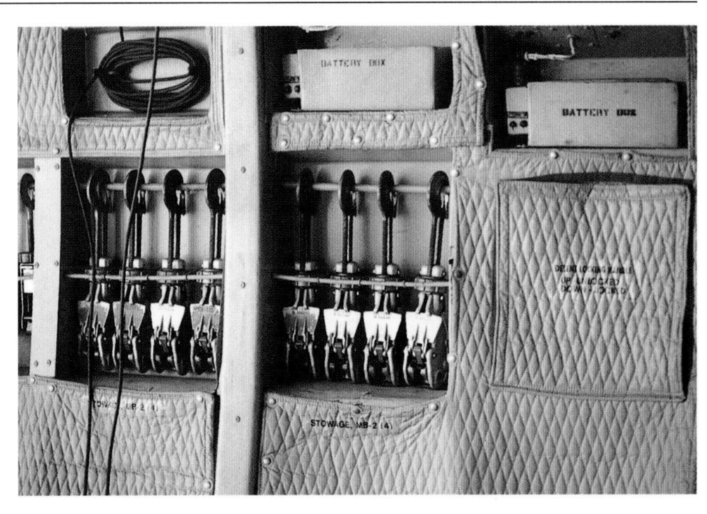

The cargo compartment walls are also used to store equipment such as cargo tie-downs and batteries. (Chris Reed)

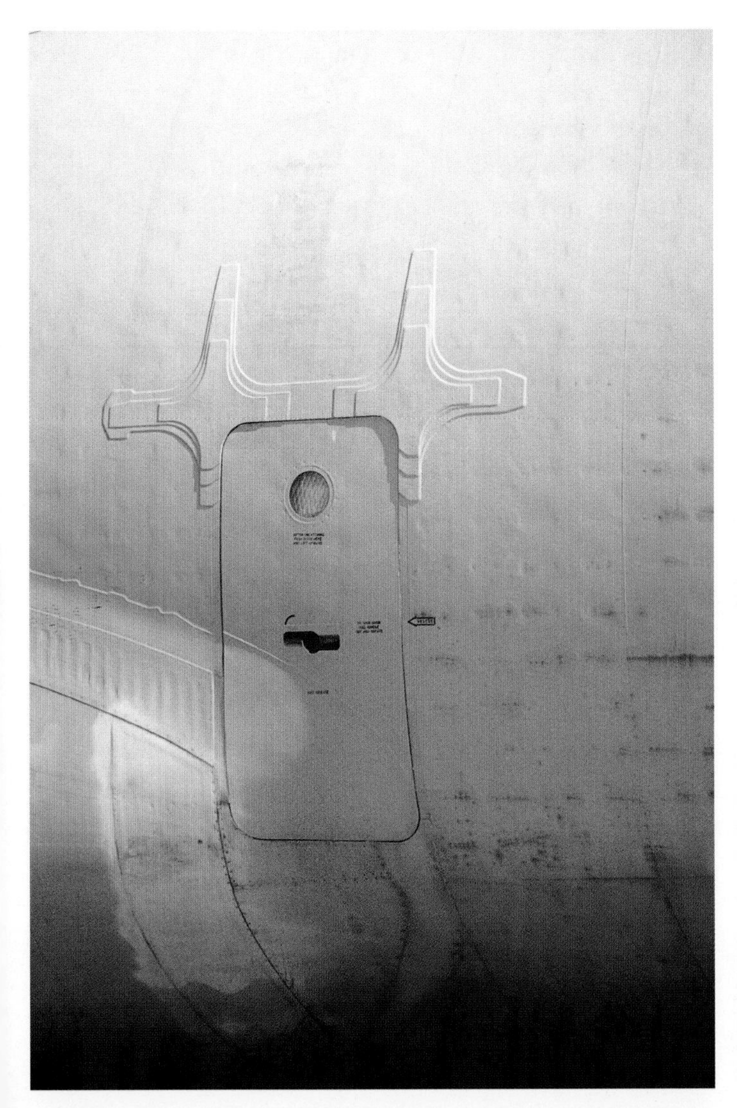

Personnel door, at the very aft end of the landing gear blister. (Chris Reed)

Side compartment for storing an engine cover. (Chris Reed)

Chapter Three

C-5B

747 versus Galaxy

Shortly before the awarding of the CX-HLS contract, Boeing had begun to seriously study using the C-5A technology towards building a heavy airliner able to carry up to 400 passengers. Designated the Model 747, the aircraft was first envisioned as being somewhat smaller than the CX-HLS designs, with a double-deck "figure eight" cross section. This gradually evolved into an aircraft that would rival the Galaxy in size, with a widebody fuselage.

Although the 747 was to predominately be a commercial aircraft, plans for a military version were very much in evidence from the early 1970s onward. The company had flown the prototype aircraft in tanker configuration in 1972, and made no secret of its desire to sell a tanker/transport to the USAF. Although first marketing the 747 as an airliner, Boeing intended to shift the aircraft over to a cargo role by the mid-1970s, as it anticipated that the flying public would be predominantly flown onboard 2707 SSTs starting in the mid-1970s.

Although the 1973 *Nickel Grass* airlift to Israel went a long way towards establishing the Galaxy's reputation after its early troubles, the operation also pointed out serious shortcomings in U.S. strategic airlift capacity. There were several possible solutions put forward, such as stretching the C-141A fleet and adding refueling receivers, and fitting the KC-135A force with modern, fuel-efficient engines. These programs were to be carried out, but there was also interest in buying new outsized aircraft. This was to lead to a face off between the giants from Marietta and Seattle.

Much of the 747's technology had sprang from the CX-HLS work, but the design was a commercial aircraft first and foremost, leading to some difficulties adapting it to a military role. The 747's configuration did not permit the easy carriage of items such as tanks, but Boeing did propose fitting the aircraft with a larger aft cargo door (135" high and 200" wide). This would be large enough to allow a pair of M60s to be brought aboard, via an air transportable side loader. A Galaxy-style folding visor nose was also feasible; tanks would not fit through the nose, but M113 APCs and light trucks could; again, a loader would be necessary.

Boeing, which had developed the Flying Boom method of aerial refueling in the late 1940s, also proposed fitting out 747s with the system, with one central boom in the conventional position under the tail, as well as a pair of remotely controlled wingtip booms. The company actually demonstrated such an aircraft, using the prototype 747 (with a single fuselage-mounted boom) to carry out "dry" refueling tests with a B-52 and other aircraft. And the practicality of using the 747 for at least some transport missions was demonstrated when a 747C owned by World Airways carried out a month-long series of evaluation missions for MAC, operating on Pacific routes out of Travis AFB, California.

Developed as a commercial aircraft, the Boeing 747 owes much of its heritage to the CX-HLS program that spawned the C-5A. Small numbers of military models have been sold, including this VIP transport example of the Japanese Air Self-Defense Force (Nick Challoner)

Galaxy 87-0031 of the 436 AW, wearing the European 1 scheme originally appplied to the C-5B fleet. By the time this photo was shot in July 1995, the new overall gray scheme had been around for several years, and the european colors are clearly showing their age. (Arno J.A.H. Cornelissen)

Lockheed, ever eager to resume Galaxy production, proposed a "C-5B" tanker/transport as competition, with a Flying Boom fitted above the aft door, along with a pair of underwing drogue pods. The basic fuel load could be supplemented by 144,000 lbs in cells under the cargo floor; altogether this tanker Galaxy would, under some conditions, be able to do the job of three KC-135As. Additionally, the TF39 engines would be replaced by later, more powerful turbofans; either the General Electric CF6-50E derivative rated at 52,500 lb thrust, or Pratt & Whitney's 52,000 lb-rated JT9D-70. A procurement figure of 100 aircraft was discussed.

As it turned out, neither the C-5 or 747 tanker/transports were to be built for the USAF, as the Advanced Tanker Cargo Aircraft (ATCA) role was met instead by a version of the McDonnell Douglas DC-10-30CF, the KC-10 Extender. Ironically, the first aircraft to be refueled by a KC-10 was a C-5A.

C-5B

Although the KC-10 and upcoming C-141B and KC-135R programs did offer some improvements in U.S. airlift capacity, by 1978-1980 MAC still desperately needed extra capability. The CX-X requirement for a new outsized transport had been around since 1974, but had not yet resulted in the acquisition of any hardware. Now driven by Soviet adventurism in Afghanistan that threatened Middle Eastern oil supplies, and fueled by increased military spending, the CX program was born. Early plans called for selecting a design by late 1980, and achieving an initial operating capability by 1986-87. The CX was intended to achieve the long-coveted capability of putting outsized loads as close to the battlefield as possible, while using a design that would take up less airfield space than a Galaxy.

The difficulties of meeting such requirements, which called for an aircraft with capabilities common to both the C-130 and C-5, were eased by the advances brought about by the Advanced Medium STOL Transport (AMST) program of several years before. AMST, which had been started in the early 1970s to provide a replacement for the C-130 in the tactical airlifter role, had been canceled in 1978, but not before a pair of competing designs from Boeing and McDonnell Douglas had been extensively test flown. Both of these would form the basis for larger CX proposals, while Lockheed was to offer a new design that owed much to both the Starlifter and Galaxy.

C-5B 85-0003 landing at Mildenhall. (Robbie Robinson)

The proposed Marietta aircraft design in many ways resembled a scaled-down Galaxy (or scaled-up Starlifter) without the visor nose. With a maximum weight of 480,000 lbs, the new design would have been powered by four Pratt & Whitney PW2037 turbofans, and would have been able to haul 130,000 lb payloads over 2,500 nautical miles unrefueled.

Boeing, whose YC-14 had used upper-surface blowing to achieve STOL performance, incorporated this technology into its CX competitor, but also fitted a third engine in a 727-style tail. The company had also studied a number of other CX configurations, including much larger aircraft in the 800,000 lb weight category.

McDonnell Douglas' aircraft was the largest of the three, with a design extrapolated from the YC-15 prototypes. There were to be significant differences, as the new aircraft was considerably larger, with sweptback rather than straight wings, and four PW2037 engines. The McDonnell Douglas design, later designated C-17, was chosen over the Lockheed and Boeing aircraft in August 1981. Similiar in overall dimensions to the C-141B, the C-17 had a much wider fuselage to permit outsized items to be carried; the C-17 would be the only aircraft aside from the C-5 able to carry the M1 tank.

With prospects for selling their new designs gone, both Lockheed and Boeing pushed their existing C-5 and 747 aircraft as interim airlifters, pending the arrival of the C-17. Lockheed in late September proposed restarting the Galaxy line for 44 aircraft, this later being amended to fifty examples. Boeing shortly afterward offered a twenty-aircraft 747 program. However, later that year, an airlift review came to the conclusion that the best solution would be to continue with the CX/C-17 development, and to buy as well an additional 44 KC-10 Extender tanker/transports.

Despite the decision in favor of the C-17, the C-5N was to get an unexpected, last-minute reprieve. In a surprise move, in January 1982, Deputy Secretary of Defense Frank Carlucci (later the Secretary of Defense) made the decision that the USAF would instead buy fifty C-5Ns, to be given the service designation C-5B. The decision was to raise a good deal of high level debate, with both the House and Senate Armed Services Committees announcing in late January that they would be holding hearings into the matter.

C-5B 85-0001 was serving as the 436 AW commander's aircraft when seen at Mildenhall on August 6, 1994. (Robbie Robinson)

60 AW "Travis" tailband on a C-5 at RAF Alconbury, September 17, 1994. (Robbie Robinson)

The attachment point between the visor nose and the weather radar can easily be distinguished. The C-5B was built with a commercially-derived radar that was much simpler (as well as cheaper and more reliable) than the set originally fitted to the C-5A. (Arno J.A.H. Cornelissen)

C-5B 87-0027 at RAF Mildenhall, May 1997. (Arno J.A.H. Cornelissen)

Supporters of the 747 kept pushing, claiming that the USAF would be better off buying immediately available excess airliners from the commercial market and having them refurbished by Boeing. However, despite the problems of the C-5A program, many in the Air Force still preferred to buy new Galaxies. A major factor in this was the C-5's roll-on/off loading capability; additionally, proponents of the C-5 pointed out that the aircraft had originally been designed for forward operations, and that if the aircraft's true capabilities were taken into account, it could use many of the same airfields as the C-17.

The Galaxy had other advantages over the newer aircraft; although the C-17 could carry outsized cargo, it could not carry as much, and could not fly as far without stopping or taking on fuel from tankers. There was also some resentment that the USAF was being pressured to bail out airlines by buying 747s that did not meet its own needs. High-level USAF antipathy to procuring the 747 was pronounced, as evidenced by MAC commander General James Allen's comments at the hearings. Allen remarked that putting the 747 in front-line service would be only a last resort, saying that he would prefer that should he be forced to take such aircraft, that they be put into the Civil Reserve Air Fleet under lease, or into Air National Guard or Reserve squadrons.

In the end, the C-5B order was to be confirmed, in December 1982. However, as part of the deal, three 747s were also to be bought.

53-56 87-0027 again, seen here in July 1996. (Nick Challoner)

87-0037 of the 436th on a gray day at Ramstein AB, Germnay, June 1998. (Arno J.A.H. Cornelissen)

As late as July 1997, this C-5B visiting Ramstein was still wearing the European 1 scheme, which had caused overheating problems during the gulf crisis years earlier. (Arno J.A.H. Cornelissen)

The Boeings were to have been designated C-19A, and it was proposed that the type would partially equip the 105 Military Airlift Group, a unit that was converting to the C-5A; apparently, it was intended to concentrate the different types of ANG strategic airlifters within a single unit. In addition to military duties, the C-19s, had they been bought, may have also been used to supplement NASA's 747 in the shuttle carrier mission. Ultimately, the C-19s were never put into service.

The C-17 program did not die with the C-5B decision, as even with the procurement of new Galaxies, there would be a shortfall in airlift capacity. Thus, the C-17 program would proceed, but it would not be until December 1985 that the order for the first prototype was actually placed, with the first flight taking place on September 15, 1991.

Although the C-5B was identified as being "identical" to the C-5A, to emphasize the low-risk nature of the program, there were to be detail differences between the variants. Of course, the updated wing design retrofitted to the C-5A was built into the B-model, but there were numerous other changes to reflect advances in materials and electronics. Improved TF39-1C engines rated at 43,000 lbs thrust each replaced the earlier -1 powerplants; other updates included electric cargo winches and the deletion of the crosswind landing capability.

The original Norden multimode radar of the C-5A, which had caused so many reliability problems, was replaced by a Bendix AN/APS-133 color weather radar, which had also been fitted on late-model C-130s. A version of the company's RDR-1FB radar, the -133 was also backfitted to C-5As beginning in 1983; a smaller radome was fitted as part of the modification. Another trouble-prone system, MADAR, was replaced with the improved MADAR II, which aside from having more memory (4 megabytes) than the original set, also has a new CRT display for the flight engineer; nearly 1,400 data points can be monitored.

After the C-5B program had begun, original plans for brakes made of beryllium were scrapped, with lighter and more efficient carbon units being substituted. And finally, the C-5A's Doppler-

As sleek as a Galaxy ever gets; C-5B 87-0041 climbs out from RAF Mildenhall in May 1997. (Arno J.A.H. Cornelissen)

87-0028 at Mildenhall, March 18, 1989. (Nick Challoner)

inertial navigation system was replaced by three Carousel 4E inertial systems.

Assembly of the first aircraft (83-1285) began in April 1984, and rollout took place on July 12, 1985. A 55-hour flight test program began on September 10, with a three hour, sixteen minute maiden flight. This first mission went well, although a seal from the left wing fell off on takeoff. Within several weeks of the first flight, the C-5 was grounded for a short time while incorrectly installed aluminum nuts were replaced by steel ones.

Even as late as the fall of 1985, the future of the C-5B program was still not entirely firm, as some within the Air Force were considering cutting the total to be bought by almost two-thirds, to 16 aircraft, to meet projected budgets. On the other hand, a year later the Congressional Budget Office, while looking at alternatives to the C-17, offered several plans, including buying up to seventy additional Galaxies. Ultimately, the original order for fifty aircraft would stand.

When the time came to introduce the C-5B into service, the 443rd at Altus was once again at the forefront, receiving the first of four aircraft on January 8, 1986. Whereas the east coast had received the first operational C-5As, the 60th MAW would be the first line unit to fly the C-5B, on August 29, 1986, with the 436th at Dover following suit soon afterwards. Construction of the last C-5B began in January 1988, and this aircraft first flew on April 1, 1989, being delivered to the USAF sixteen days later to close out the program.

86-0025 at Mildenhall, May 1996. (Nick Challoner)

87-0045, late July 1999. (Nick Challoner)

Iran

Once a close ally of the U.S., and almost an operator of the Galaxy itself, Iran descended into revolution and chaos in 1979, setting the stage for two decades of animosity and occasional outright hostilities between America and the Middle Eastern state. C-5s were involved in Iranian operations from early on; along with C-141s, Galaxies were used during the revolution to evacuate U.S. military personnel from the capitol of Tehran to safety at several European bases.

By the late 1980s, the tension between Iran and the U.S. was to come to a head. Fearful of the Persian state gaining power over the whole gulf region, the U.S. began a tacit backing of the regime of Saddam Hussein in his war with Khomeini, and went so far as to reflag Kuwaiti oil tankers, putting the ships under American protection against attack from Iranian aircraft or ships.

Operation Earnest Will got off to a bad start, with the tanker S.S. *Bridgeton* hitting a mine on the first trip up the gulf. With only a token force of minesweeping ships, none of which could reach the gulf quickly, the USN had to once again deploy its RH-53D MCM helos to the area. Four C-5A missions brought eight RH-53Ds to Diego Garcia, where they were put aboard the USS *Guadalcanal*, the amphibious assault ship having unloaded a potion of its regular air wing to make room. This was not the first time that C-5As had assisted in the deployment of MCM helos; in 1984, RH-53s had been taken by Galaxy to Rota, Spain, from where they proceeded under their own power across the Mediterranean for minesweeping activities in the Red Sea.

Galaxies for the Guard and Reserves

While associate units had been flying the Galaxy since 1973, by the mid-1980s the pending reequipment of the 60th and 436th MAWs with B-models meant that for the first time C-5As would be available to equip the Air National Guard and Air Force Reserve, which had lost their last strategic airlift units with the retirement of the C-124 more than a decade before.

The 68th MAS of the 433rd MAW at Kelly AFB, Texas (formerly the 68th TAS/433rd TAW, flying C-130s), was the first Reserve unit to convert, receiving its first C-5A in December 1984.

(Andy Spagna)

(Andy Spagna)

(Andy Spagna)

The only Air Guard squadron, the 137th MAS/105th MAG at White Plains, made the enormous transition from flying the Cessna O-2 Skymaster in the FAC role to operating strategic airlifters, taking charge of the first of its new aircraft in the summer of 1985. And finally, the Reserve's 337th MAS/439th MAW traded in their older model C-130s for the Galaxy in late 1987.

Central & South America

Whereas much of the superpower tension of the 1980s revolved around Europe, Central America was another Cold War crisis point, with the U.S. faced with increasing communist influence not far south of its own borders. Although the involvement of U.S. forces did not lead to the feared ensnarement into a "second Vietnam" forecast by some, the American military was nonetheless deeply involved in the region throughout the decade.

Notable among the shows of force was *Operation Golden Pheasant*, the rapid deployment of 3,600 members of the 82nd Airborne and 7th Infantry Divisions to Honduras in March 1988. Honduras, which hosted bases of the Contra rebels fighting the Sandinista government of neighboring Nicaragua, had requested an American military presence following strikes by the Nicaraguan army across the border. MAC shortly thereafter began the *Golden Pheasant* airlift, 55 missions being flown by C-5s and C-141s.

Leftist guerrillas and Communist governments were not the only problems faced by the U.S. in Central and South America, as efforts were also underway to combat narcotics traffic originating in the region. On September 5, 1989, during a period of increased drug-related violence in Columbia, a C-5 arrived in the capital city of Bogota carrying surplus Huey helicopters earmarked for use in the Columbian government's counterdrug campaign.

Panama

By far the largest U.S. military operation in Central America during the 1980s came in the last days of the decade, when on December 20, 1989, the *Operation Just Cause* invasion of Panama was launched. In many ways a proving ground for the technologies, tactics, and doctrine that would later be applied against Iraq, *Just Cause* came about as tensions heightened between the U.S. and Panamanian leader General Manuel Noriega, who had been wanted since 1988 on U.S. drug charges. The general had also deposed the Panamanian president, nullified the results of a presidential election, and put down several coup attempts.

By the spring of 1989, the U.S. and Noriega were on a collision course, with American troops being deployed to Panama under *Operation Nimrod Dancer* to give them experience in the country they would be tasked with subduing, as well as to conduct freedom of movement exercises.

(Andy Spagna)

86-0013 of the 9AS/436 AW landing at London, Ontario on June 5, 1992. In the foreground is AV-8B Harrier 163199 "WH-15". (Tim Doherty)

The longtime U.S. presence in Panama was a godsend to planners, as there would be forces and facilities already in place when the operation began, service people were familiar with the area, and preparation for the invasion could be disguised as normal training.

The breaking point was finally reached in December 1989, when an American serviceman was shot and killed by the Panamanian Defense Forces, and others were held against their will. The previous day, Noriega's National Assembly had declared that Panama was at war with the U.S.

The invasion began just before 1:00 AM local time on December 20; hours earlier C-130s and C-141s had left California and North Carolina carrying members of the 7th Light Infantry and 82nd Airborne Divisions. C-5s had already brought in a small force of Army AH-64 Apaches to conduct precision strikes; the Panama operation would be the Apache's baptism of fire.

Howard AFB was a center for airland operations of transports arriving from the U.S., as well as a base for OA-37s, A-7Ds, and special operations aircraft. The airport at Toorijos-Tocumen would also be used, after being seized through parachute assaults by Army Rangers and paratroopers from the 82nd. Altogether, C-130s and Starlifters dropped almost 4,000 paratroopers, making *Just Cause* the largest nighttime airborne combat operation since the D-Day drops of June 1944.

C-5s and Starlifters continued to bring in follow-on troops from the 7th LID and other units, but large-scale resistance from the PDF and Noreiega's "Dignity Battalions" was silenced well before Christmas, and Noriega himself was arrested in early January.

87-0028 of the 60 AW, seen at Dyess AFB, Texas in late 1999. In the background are B-1B Lancers, homebased at Dyess. (Tim Doherty)

C-5B 86-0042, 60th AW, July 11, 1993. (Masanori Ogawa)

Pacer Snow

The threat to large transports from shoulder-fired SAMs, such as the SA-7 Strela, had been present since at least the early 1970s, but serious attention to countering this problem did not take place until well into the 1980s. During that decade, the Soviets lost a large number of aircraft in Afghanistan to U.S.-supplied FIM-92 Stingers, leading to crash efforts to fit chaff and flare dispensers to many types of aircraft, including transports. The U.S. was actually behind in this field, but the rise in terrorism and the proliferation of light SAMs was heightening the threat; it was recognized that had Stinger-class missiles been available in Panama they could have played havoc with the invasion there. This led to programs to fit transports operating into high-threat areas with missile warning systems and countermeasures. The effort for the C-5, codenamed *Pacer Snow*, entailed fitting the Loral AN/AAR-47 missile warning system to detect incoming SAMs, and AN/ALE-40 flare dispensers to decoy them.

Deep Freeze

Although much of the fixed-wing airlift support for the National Science Foundation's Deep Freeze operations have come from LC-130s operated by the U.S. Navy and later the Air National Guard, larger aircraft have also visited the ice-bound continent, including USAF C-124s and C-141s. Although not fitted with skis for operations deep into Antarctica, Starlifters could nonetheless operate from the blue-ice runway at McMurdo Sound, during the period which this is useable. Although more capable than the LC-130s, the C-141s were still not able to handle some outsized loads (such as helicopters) in assembled form, a critical capability when work must be done outside in one of the harshest environments on the planet. Thus, on October 4, 1989, the 60th MAW conducted the first ever Galaxy mission to Antarctica, staging from Christchurch, New Zealand.

The largest aircraft to operate in the southern polar region, the C-5 requires some consideration when at McMurdo; despite the runway ice being nearly eight feet thick, the aircraft is ready to taxi at all times in case the surface begins to give way. And, as with all aviation activities on the continent, crews must take steps to make sure that hydraulic and other systems are not degraded or frozen altogether by the intense cold.

Desert Shield/Storm

As the 1990s dawned, American armed forces were for the most part at the zenith of their post-Vietnam strength, having benefited from the increased defense budgets of the early Reagan years, and the introduction of a whole new generation of systems. However, with the ongoing process of reform in the then-Soviet Union and the breakup of the Warsaw Pact, the U.S. military was facing an uncertain future as its primary opponent was disappearing, and domestic budget pressure had prompted the beginning of what would

C-5B 86-0022, 60th AW, May 23, 1993. (Masanori Ogawa)

be a major force drawdown. Few could have predicted that while this was going on, the country was also shortly to become embroiled in its largest conflict in decades.

Since early in the summer, Iraq, under the leadership of Saddam Hussein, had been making bellicose statements regarding neighboring oil-rich Kuwait. Such threats and posturing were nothing new; Britain had rushed forces to the Gulf in 1961 to deter an invasion, and there were further scares in the 1970s. It was thought that Hussein would, at the most, try to grab some disputed territory. This notion was disabused early in the morning hours of August 2, when Iraqi armor, backed by special operations units and airpower, poured across the Kuwaiti border, and soon fighting was taking place in the capital of Kuwait City. Some elements of the Kuwaiti armed forces put up a spirited fight, but the issue was never in doubt, and shortly Hussein controlled a sizable portion of the world's known oil reserves, and was in a perfect position to move across the Kuwaiti border into Saudi Arabia. The Saudi government formally asked for U.S. military help on August 5, and President Bush responded by authorizing the *Operation Desert Shield* deployment, the largest such movement of U.S. forces since Vietnam.

The American military had been planning means of getting combat power into the Gulf region for years, but for years these concerns had been mostly centered on countering Soviet moves in the region. Indeed, it was this kind of threat that prompted the creation of the Rapid Deployment Force in 1979, which evolved into the U.S. Central Command in 1983. Fortuitously, Central Command had, just weeks before the Iraqi invasion, run the *Internal Look* series of wargames that dealt with a similar scenario.

Like other parts of the post-Vietnam military, the USAF's airlifter force relied heavily on Guard and Reserve crews and aircraft, and from the outset of the gulf crisis ANG and Reserve personnel were flying missions on a volunteer basis, even before their units were mobilized.

However, the mission of taking such immense amounts of material overseas meant that a formal call-up of Reserve and Air Guard units could not be put off. On August 22, President Bush signed a federalization order for selected units, the first such instance since the Vietnam War. With the exception of a single ANG RF-4C squadron, all were transport or tanker units, including the Air Guards 137th MAS/105th MAG from Stewart, New York, and the 68th and 33rd Squadrons of the Air Force Reserves 439th MAW at Westover, Massachusetts. This initial call-up entailed the units being on active duty for ninety days, but many reservists and guardspeople would nonetheless still be in service well into 1991.

During the early days of *Desert Shield*, a priority was getting anti-tank systems to Saudi Arabia as quickly as possible. Deployment of any significant number of U.S. M1 Abrams MBTs was tied to the arrival of fast sealift ships, but a number of Galaxy flights carrying armor were made, footage of which was given much airtime to foster the impression among Iraqi military leaders that there were more U.S. tanks in-theater than were actually present.

Helicopters were a prime cargo; one such tasking was the airlift of the 229th Aviation Regiment's 2nd Battalion; in just over a week ten Galaxy sorties and an equal number of Starlifter missions took the battalion's nearly three dozen helicopters from Fort Benning, Georgia, to the theater. Galaxies also brought over other, more unusual aircraft, such as MH-60 Pave Hawks and MH-6 Little Birds of the 160th SOAR to conduct special operations in the theater.

Aside from the forces and weapons themselves, munitions shipments were also necessary, and in the early days of the operation C-5s carried in armament supplies for the deployed units from stocks in western Europe.

Sadly, there was to be a fatal Galaxy crash during Desert Shield, the only instance of an airlifter being lost during the operation. On August 29, C-5A 68-0228 crashed just after take-off from Ramstein AB, Germany. Flown by reservists of the 433rd, but belonging to the 60th MAW, -0228 was carrying a crew of 17 when it went down; four survived the crash, which was attributed to a thrust reverser malfunction.

Although Iraq's MiGs were unable to venture into Saudi airspace, since it was guarded by Coalition interceptors and Patriot and Hawk SAM batteries, MAC and CRAF crews still had to face the possibility of enemy action, including strikes by *Scud* ballistic missiles carrying biological or chemical warheads. This latter threat led to hurried efforts to locate and ship enough NBC containment garments for airlifter crews.

By October 1990, the U.S. finally had sufficient forces in the theater to adequately counter an Iraqi attack, as well as having a limited offensive capability. But the costs of maintaining such a force in the Saudi desert indefinitely were prohibitive from military, political, and financial standpoints. In early November, President Bush announced that further deployments would be made to give CENTCOM and the Coalition the means to forcibly evict Iraqi forces from Kuwait. Airlift forces, which had transitioned to a support role, once again began transporting units to Southwest Asia.

86-0021, 60th AMW, July 18, 1998. (Masanori Ogawa)

The official end of the Desert Storm war in early March 1991, which left Hussein in power and a sizable portion of his military intact, really only marked the finale of the first phase of a U.S.-Iraqi conflict that would span the 1990s. In the longer term, a greatly increased U.S. military presence in the area would be on hand to keep Hussein contained. And periodically, Iraqi moves toward Kuwait, its own Kurdish population, or noncooperation with UN weapons inspectors would lead to renewed deployments and clashes. Airlifters would also support the operations over the northern and southern no-fly zones, and the Army's *Intrinsic Action* deployment/training rotations in Kuwait.

Air Mobility Command

In the early post-*Desert Storm* era, the USAF was to undergo major organizational changes to help the services meet new challenges with fewer available resources than during the Cold War. As of June 1, 1992, Military Airlift Command was no more; now, strategic airlift and most refueling tankers would come under the control of Air Mobility Command, headquartered like its predecessor at Scott AFB, Illinois.

The creation of AMC would not be the only change to effect airlifter forces, as the 443rd at Altus, having already been redesignated as an Airlift Wing in 1991, was replaced on October 1, 1992, by the 97th Air Mobility Wing. In July of the following year, the 97th would shift to the control of the new Air Education and Training Command. The 60th MAW became an Airlift Wing in November 1991, and underwent further redesignation as an Air Mobility Wing in October 1994 after having taken charge of a squadron of KC-10 tankers.

Despite the end of the Cold War, the situation in many parts of the world was actually less stable than before, with old allegiances gone and ancient hostilities working themselves back to the surface. At the same time, the U.S. military, bereft of its traditional Soviet adversary, nonetheless found itself occupied with overseas operations, bringing relief to strife-torn countries, and often being used to impose peace between warring parties.

87-0032, 60th AW, September 15, 1996. (Masanori Ogawa)

Beginning in late July 1994, the USAF launched *Operation Support Hope*, a humanitarian relief effort for millions of Rwandan Hutus that had been forced to flee to Zaire and elsewhere. USAFE C-130s flew the greatest number of sorties, but AMC C-5s and Starlifters made almost 400 missions, taking on fuel when necessary from 100th ARW KC-135Rs operating out of Mildenhall, England.

Zaire itself would be the scene of much strife several years later in 1997, prompting *Operation Guardian Retrieval*, the deployment of forces to prepare for an evacuation of U.S. citizens and other third-party nationals. C-5s were part of the airlift force, supported by KC-135s flying out of Moron, Spain. Almost two decades earlier, in 1978, C-5s and C-141s had also been active in the area during a rebel offensive in Zaire, helping to evacuate westerners and supporting Belgian and French forces deployed to the country.

Somalia

Somalia, which had fallen into civil war and anarchy in 1991, was another country that saw American military intervention to restore order and supply aid to those in need. Started with the highest hopes, the Somalia intervention eventually resulted in the ensnarement of U.S. forces into open combat with Somali factions, and the largest single American military defeat since Vietnam.

American activities actually began in August 1992, when C-130s deployed to Kenya began operating into Somalia as part of *Operation Provide Relief*. Despite international efforts, the situation continued to deteriorate into the fall, as rival clans disrupted the flow of supplies and appropriated much material for their own use.

In order to ensure the ability of relief organizations to operate safely and distribute supplies as needed, *Operation Restore Hope* was launched in December 1992. The initial stage had U.S. Marines landing at Mogadishu to secure the airport, allowing AMC transports to bring in troops, vehicles, and equipment. As was the case with other operations, C-5 and Starlifter crews flying into Somalia had to cope with limited navigation aids, lack of apron space, fuel supplies, and other ground support. Getting supplies from Mogadishu out into the field entailed transhipping cargoes aboard C-130s, which could operate from the even more marginal secondary airfields throughout the country.

Despite its initial success, the Somalian intervention did not cure the basic problem of heavily-armed clans jockeying for control of territory and power. The tone of the operation began to change in June 1993, when Pakistani soldiers were ambushed and killed, an action attributed to the forces of warlord Muhammed Farah Aideed. Operations aimed at capturing Aideed began, and in August 1993 C-5s brought into Mogadishu Task Force Ranger, composed of elements of the Delta Force counterterrorism unit, TF160 aviation assets, and Army Rangers to aid in the hunt. These were engaged in a ferocious, hours-long firefight on October 3-4, after a raid to capture Aideed lieutenants had gone wrong when a UH-60 was downed by an RPG. Without means of extraction, U.S. personnel suffered 18 dead and 78 wounded. The disastrous raid led to a major reinforcement of U.S. forces in Somalia, which Air Mobility Command assisted with. Despite this deployment, the Mogadishu disaster had sealed the fate of the U.S. mission, and the last U.S. personnel left in the spring of 1994, the final flight out of Mogadishu being covered by U.S. Marines, who then left for waiting ships.

Project Sapphire

Among the new challenges to U.S. national security in the post Cold War world was the threat of nuclear proliferation, fueled by the dispersal of weapons, material, and engineers from the former Soviet Union. Efforts were made to safeguard ex-Soviet weapons and material from sale or theft; one such operation, *Project Sapphire*, took place under great secrecy in the former Soviet republic of Kazakhstan, which possessed 600 kg of enriched weapons-grade uranium. Unneeded by Kazakhstan, this material could have been used to build as many as 36 bombs, leading to the joint mission that involved a quartet of Galaxies.

After the fuel had been prepared for transport to the U.S., two C-5s flew the material to Dover AFB nonstop, from where it was brought to Oak Ridge, Tennessee, for use as reactor fuel. Two other Galaxies also participated, one as a spare and the other as a support aircraft.

Non Developmental Airlifter Aircraft

Surprisingly, considering the controversy stirred by the C-5A/B programs, the Galaxy was almost put back into production yet again, in the mid-1990s. Slowed down by the end of the Cold War, the C-17 program had been scaled back drastically. Originally, plans had called for 210 aircraft to be bought, but even before the first flight this figure had been cut nearly in half, to 120. Suffering weight problems and cost overruns, the program was put on probation in early 1994, with procurement to be limited to only the first forty aircraft until the program showed signs of improvement, or an alternative could be found. With the C-17 looking to be a likely candidate for termination, a hunt was begun for a non-developmental airlifter that could act as a supplement or replacement for further Globemaster orders.

Lockheed's candidate was the C-5D, essentially a B-model with new avionics and CF6-80C2 engines. A glass cockpit derived from Lockheed's C-130J work would ease the flight crews' workload, while more modern systems and the more powerful engines would improve the Galaxy's reliability rates.

McDonnell Douglas did not take the threat to the Globemaster program lightly. The company also attempted to hedge its bets by offering the MD-17, a proposed commercial freighter version of the Globemaster with many of the more expensive military features deleted to cut costs, as well as a military version of the MD-11 airliner, which would have had some level of compatibility with the USAF's KC-10 fleet.

The other serious competitor was the Galaxy's old rival, the Boeing 747. Boeing now proposed a military freighter version of its latest 747-400; designated C-33A, the Boeing aircraft would be substantially cheaper than the Globemaster (at around $150 million a copy) and have a greater unrefueled range. Adaptations would include widening the cargo door, making provisions for relief crews on longer missions, and beefing up the cargo floor for carrying extra-heavy payloads. There were other contenders; several firms proposed to refit existing 747s or DC-10s for military duties.

This latest airlift review ended on November 4, 1995, when Deputy Secretary of Defense John White announced that the analysis had shown that the best solution would be to buy the full complement of C-17s, and that McDonnell Douglas had made much progress in turning the program around. Additionally, a month long series of operational tests in the summer of 1995 had shown that the Globemaster could sustain high availability rates, even under simulated wartime conditions.

Although a final decision of buying the C-33 was put off until a review of the CRAF program could be completed, this effectively put an end to C-33 procurement plans, as thanks to the large-scale purchase of Globemasters, the number of 747s needed would be so small as not to be cost-effective. Despite being able to carry an estimated three-quarters of all air-transportable items, the 747 could not take the very critical warfighting material that would be necessary in the first days of an emergency; AMC planners had to plan for airlift operations into active war zones, unlike the tense peace that held during *Desert Shield*. The C-17 could carry Bradley IFVs, Patriot SAM/ABM batteries, and M1 MBTs, while also using austere fields.

Haiti

Haiti was another of the world's troublespots in the early 1990s, with a military group seizing control from President Aristide in 1991. The country's infrastructure crumbled, and a humanitarian crisis developed. Faced with thousands of Haitians attempting to escape

87-0034, 60th AW, January 29, 1994. (Masanori Ogawa)

the country on improvised rafts, the U.S. began planning to depose the Haitian military, restore Aristide, and rebuild Haiti.

On September 19, 1994, during negotiations between the junta, former President Carter, and retired General Colin Powell, a diplomatic solution to the crisis was found. This breakthrough came at almost the last minute, as the U.S. military had already begun the initial stages of *Operation Restore Democracy*, the invasion of Haiti and the forcible removal of the junta. In what would have been the largest airborne operation since World War II, 60 C-130s carrying nearly 4,000 paratroopers and equipment had left Pope AFB, North Carolina, before receiving word that there would not be a forcible entry into Haiti.

Restore Democracy was canceled, but the *Uphold Democracy* occupation and rebuilding of Haiti was underway by the next day.

Whereas large transports could operate from the airport at the capitol of Port-au-Prince, the country's other major airfiled at Cap Hatien could not handle heavy aircraft, so missions were also flown into the naval air station at Roosevelt Roads, Puerto Rico, with C-130s then transhipping cargoes from "Rosy Roads" into Haiti.

The multinational nature of the operation also meant that AMC transports had to make overseas flights to take peacekeepers from various nations to Haiti.

C-5M Modernization Program

While the first decade of the Galaxy's life had been overshadowed by the wing fatigue problems, by the 1990s the C-5 looked to have a long life ahead of it, at least from a structural standpoint. However, serious shortcomings in Galaxy readiness rates were in evidence, driven by the TF39 engines and antiquated 1960s-era avionics.

Aside from having an updated cockpit, the C-5M would also be able to take advantage of the development of more powerful and efficient engines. Continuing the Galaxy's history with General Electric powerplants, Lockheed announced in 1997 that the engine chosen for the C-5M program was GE's CF6-80C2, which would dramatically improve the C-5M's performance, shortening the take-off roll by 30% and improving time to climb by 58%, while at the same time improving fuel efficiency. With the proven commercial engines the C-5M would also enjoy increased reliability and easier support; by 1997 GE had delivered over 2,000 -80C2s for MD-11s, 747s, and 767s. Military applications included the F103-GE-102 version for the VC-25A "Air Force One" special air mission transport, and the four E-767 AWACS of the Japanese Air Self Defense Force.

Also under consideration at the time was a proposal to replace the C-5 entirely with a C-17 variant stretched by as much as forty feet. This would increase the Globemaster's volume, but there would be tradeoffs—for example, the C-17's rough field capability would be diminished, and the type's F117 engines would need to be replaced. In April 1999 Boeing floated a plan for an additional sixty C-17s to be bought; these would apparently be standard length aircraft, albeit with additional fuel tankage.

Desert Fox

Among the most serious post Desert Storm clashes between the U.S. and Iraq was *Operation Desert Fox,* a four day series of airstrikes and missile attacks in December 1998 following the final withdrawal of UN arms inspectors. Tensions had been running high for more than a year over the inspections issue and Iraqi threats

87-0042, 60th AW, February 6, 1994. (Masanori Ogawa)

against U.S. U-2s, and AMC had conducted several airlift operations to support deployments to the gulf: *Phoenix Scorpion I* (November 1997), *Phoenix Scorpion II* (February 1998), and *Phoenix Scorpion III* (November 1998).

The end of *Desert Fox* in no way resulted in a lessening of tensions between the coalition and Iraq, as Saddam's government declared that it would no longer recognize the northern and southern "no fly" zones. This set the stage for a virtually daily series of Iraqi provocations in the zones and corresponding Coalition reprisals that would last well into 1999.

Many of the sorties for this simmering air war were flown out of Incirlik AB, Turkey, and following Iraqi threats to strike at Coalition bases, Turkey, in January 1999 requested the deployment of Patriot missiles to defend against possible Scud attacks. This was approved, and C-5s together with C-17s brought Patriot batteries from the 69th Air Defense Artillery Brigade in Germany to Incirlik.

C-5 Avionics Modernization Program

To date, the full C-5M program has not been approved by USAF, but on January 22, 1999, the service did embark on the Avionics Modernization Program, which will replace older systems with modern commercially-derived avionics. Using the benefits of its C-130J experience, Lockheed Martin is to fit the Galaxy force with "glass" cockpits incorporating six color Multi-Function Displays to replace conventional instrumentation, a Terminal Collision Avoidance System, a digital flight control system, and a communications/navigation system that meets international requirements.

Modification and testing of a pair of C-5 AMP prototypes was scheduled to take place in 2001-2002; although these first retrofits are to be undertaken at Marietta, all others will be done at C-5 bases. Still under consideration are further plans for new engines and other structural improvements. The CF6-80C2 engine announced for the C-5M is still a contender, as are other commercial powerplants, such as the Rolls-Royce Trent 500 and RB211-535E4, as well as the Pratt & Whitney PW4000.

Operation Allied Force

Continuing the seemingly nonstop operations that had preoccupied MAC—and later AMC—for the whole of the 1990s, airlift forces in the spring of 1999 were tasked with supporting NATO's war against Serbia, *Operation Allied Force*. Tensions between the western alliance and Serbia had been escalating for some time, driven by ethnic cleansing in the Yugoslavian province of Kosovo. As war with Serbia drew near in late February 1999, the USAF sent seven B-52Hs to RAF Fairford in the U.K., a move that was supported by two C-5Bs from Travis, 84-0060 and 87-0028.

As NATO airstrikes began, Serbian forces forced hundreds of thousands of Kosovars out of Kosovo and into neighboring Macedonia, Albania, and Montenegro, leading to the largest European humanitarian crisis since World War II. Airlifters were pressed into service to deliver much-needed food and supplies into the Balkans, but overcrowded conditions and poor facilities kept C-5s and 747s from operating directly into the region, with the larger aircraft having to offload supplies at Ancona, Italy, from where they were loaded into C-130s for the trip across the Adriatic. C-5s brought in tents to fill the rapidly swelling refugee camps, rations to feed the newly homeless, as well as unloading equipment and vehicles to facilitate the airlift itself.

As the air war escalated, NATO aircraft reinforcements continued to flow into Europe; Galaxy support of these deployments included a heavyweight flight from Holloman AFB to support the sending of an additional 13 F-117s to Germany and Italy. On April 13 a C-5 (callsign REACH 7041) made the trip from Elmendorf AFB, Alaska, to RAF Lakenheath, U.K., carrying support equipment for the F-15Cs of the 3rd Wing that would make the flight the following day. And the deployment of nearly the entire 171st ARW, the Pennsylvania ANG's "superwing" of KC-135Es, necessitated a number of C-5 operations from the 171st's base at Pittsburgh to ferry personnel and equipment.

NATO airstrikes ended in June 1999, but the Serbian withdrawal from Kosovo did not mark the end of the coalition's activities. NATO members were to maintain the peace within the province. With reports of widespread atrocities, efforts were made to document forensic evidence of crimes against humanity. Among the U.S. contributions was the dispatch of an FBI team on June 21, the agents and their equipment being carried over aboard a Galaxy from Dover.

VIP Support

While the 89 Airlift Wing's substantial fleet of VIP-configured transports have the mission of transporting the President and other high-level "distinguished visitors," other far less glamorous airlifters are also used to support these missions. Under the codename *Phoenix Banner*, C-141s and C-5s provide logistical support, including the transport of limousines, Secret Service Vehicles, and communications gear. Some Presidential trips are preceded by C-5 missions carrying VH-3D or VH-60 helos on HMX-1 for transportation from the destination airfield. Similar missions in support of the vice president are codenamed *Phoenix Silver*.

87-0040, 60th AMW, December 9, 1999. (Masanori Ogawa)

Chapter Four

Future Heavy Lifters

In 1967, even before the first C-5A had flown, Lockheed was already considering possible alternative roles for the basic design.

ICBM Launcher

From the outset of the age of the ballistic missile, the Air Force had been interested in combining this new weapon, providing an airborne launcher that could not easily be targeted, while allowing the aircraft to strike targets from far beyond the range of air defenses. However, plans for nuclear-powered missile launchers came to naught, and the Skybolt ALBM for the B-52 was canceled in the early 1960s.

Renewed USAF interest in such a system was apparent from at least 1969-1970, when the service began to be concerned over the Navy's ULMS project, which would eventually result in the Trident submarine launched ballistic missile system. Fearing that the Navy project would garner increasing amounts of an ever-shrinking defense budget, to the detriment of Air Force strategic programs, the USAF looked into the possibility of putting ICBMs on large airborne platforms to provide an equivalent capability.

By the early 1970s, there was an increased need for such a program, as the accuracy and throw-weight of Soviet ICBMs was increasing, putting at risk USAF silo-based Minuteman and Titan missiles. Although most later plans called for carrying missiles internally, Lockheed also looked at fitting the Galaxy with a weapons bay or dropping the missiles from wing stations.

The Galaxy's ability to carry and fire an ICBM was demonstrated on October 24, 1974. Flying off the coast of southern California, a C-5A carrying a 78,000 lb LGM-30A Minuteman I opened its rear doors, and drogue chutes successfully extracted the missile and its 8,000 lb carriage structure. At 8,000 feet the first stage was ignited; as a full-range test shot was not planned, the engines were shut down after ten seconds, the missile by then having passed 20,000 feet. The test showed that air-launched ICBMs were feasible, but also served another purpose, as it occurred while arms

Looking more like a futuristic bomber than a transport, Lockheed Martin's Blended Wing-Body transport concept could spawn a "stealthy" replacement for MC-130 special operations aircaft. (Lockheed Martin)

Twin-engine version of Lockheed-Martin's Box-Wing transport; by 1999 the company had flown a radio-controlled model of similiar configuration. (Lockheed Martin)

talks were underway in Moscow. Coming several days after a pair of Soviet ICBM tests, it helped show that the U.S. was not behind in the field of mobile missiles.

No further moves towards an airmobile Minuteman force were undertaken, as most attention was focused on defining a new ICBM, the Missile Experimental, or MX, which would enter service a decade later as the LGM-118 Peacekeeper. As part of the MX studies, and faced with the growing threat of ever more accurate and powerful Soviet ICBMs, the USAF was by 1978-79 examining numerous schemes for survivable basing, including airmobility.

Planning for the MX centered around a 190,000 lb, 92-inch diameter weapon; both the Galaxy and 747 could carry a pair of these weapons, and thus Lockheed and Boeing dusted off their concepts for ALBM aircraft. Boeing could also offer its YC-14 AMST design for the role; although the basic aircraft could not accommodate an ICBM the size of the M-X, a stretched version could, as could a similarly scaled-up version of McDonnell Douglas' YC-15.

During the intense debate over MX basing that took place in the early years of the Reagan Administration, then Secretary of Defense Weinberger proposed that 100 missiles be deployed in ALBM mode aboard C-5s, pending the arrival of a purpose-designed launch aircraft later in the decade, at which time the Galaxies would be shifted to the airlift role. Concepts for a type designed for the mission included an aircraft with an internal weapons bay, high aspect wings nearly 400 feet in span, and turboprop or advanced diesel engines. Boeing also had several designs using technology extrapolated from the company's 747 work, all having turbofan engines, high-mounted unswept wings, and a T-tail.

Ultimately, airmobility for the MX was ruled out and deployment centered around putting 50 LGM-118s in converted Minuteman silos, follow-on plans for fifty rail-mobile missiles being can-

Scaled-up four-engined Box Wing, in commercial colors. (Lockheed Martin)

Military Box Wing; despite the unorthodox wing structure, the fuselage clearly shows the heritage of several decades of Lockheed transport design experience. (Lockheed Martin)

USAF tanker version, with remotely controlled wing-mounted refueling booms. (Lockheed Martin)

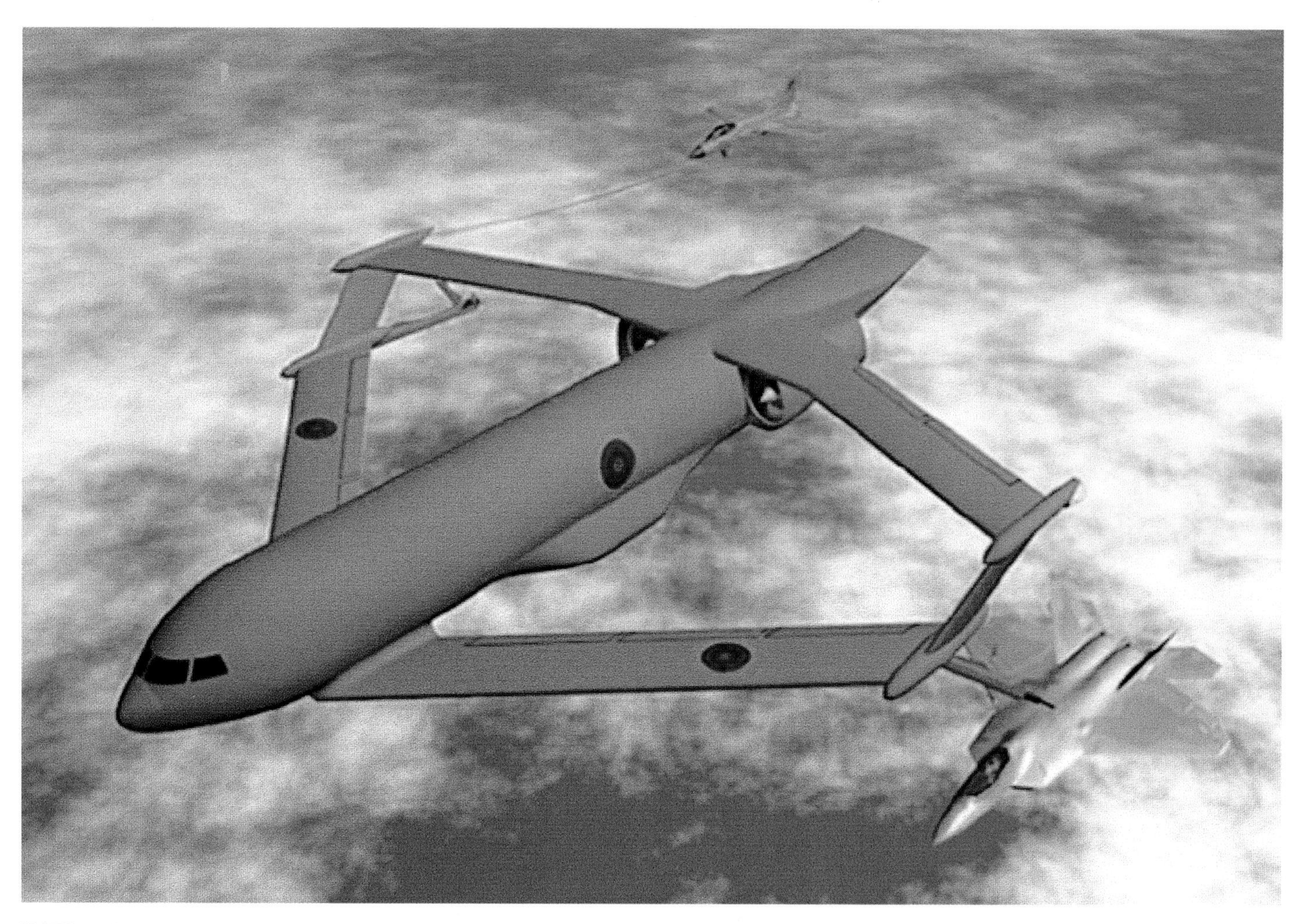

NATO tanker-transport model, with both boom and drogue refueling gear. (Lockheed Martin)

Ilyushin's Il-76 Candid freighter closely resembled the slightly earlier C-141, but was larger, and really represents an intermediate design between Starlifter and Galaxy-class aircraft. While plans for specialized C-141 and C-5 derivatives generally did not amount to much, the Candid has formed the basis for numerous variants, including the Il-78 Midas tanker model seen here. (Simon N.J. Edwards)

celed. Aircraft basing was also considered for the MGM-118 Small ICBM before the program was terminated.

Nuclear-Powered Transport

The idea of using nuclear energy as a power source for aircraft dates back to the dawn of the nuclear age, and began to receive serious consideration from 1946 on. Nuclear turbojets using the heat generated by a reactor to combust fuel held out the promise of greatly extended range and endurance, although open-cycle designs would also produce exhaust contaminated with fission by-products.

The USAF was quite interested, envisioning a long-endurance missile platform that could stay aloft for weeks without tanker support. In particular, one depiction for a tanker/transport showed a design that bore a striking resemblance to the C-133, with high-mounted unswept wings and nuclear turboprop engines.

Enthusiasm for nuclear aircraft eventually diminished; there would be massive problems in the event of a crash, shielding the crew was an enormous technical challenge, and even routine servicing would be difficult. Reports of Soviet advances in the field were no more than hyperbole, and the effort was draining funds from ICBM programs which promised more immediate results.

The A-Plane concept did not die entirely, as by the late 1960s some saw the Galaxy as the ideal platform for a testbed incorporating improved reactor technology, in advance of an entirely new and larger purpose-built design. The "Nuclear Galaxy" would have had a reactor in a sealed compartment, retaining cargo space forward and aft. The engine type was not specified, but would have been capable of conventional operations during takeoff and landing.

The USAF was not the only American service to desire a nuclear-powered aircraft, as the U.S. Navy wanted an "A-Plane" that could conduct very long endurance ASW flights and other missions; one plan in the late 1950s would have combined an airframe based on the Martin P6M Seamaster jet seaplane with nuclear engines. Hoping to attract Navy interest (and eventual orders), Lockheed proposed a new nuclear flying boat design, derived from the Galaxy. Such a redesign was not unprecedented, as Lockheed had detailed plans for a similar, albeit conventionally powered derivative of the C-130, and decades earlier Convair had examined the possibility of converting the XC-99 into a flying boat.

An-124 UR-82027 of Antonov/Air Foyle at RAF Fairford, immediately following a six-week total refurbishment. (Terry Lee/Aviation Consultants Ltd, UK)

Aside from the low-mounted tailplanes, the Condor is quite similiar in general configuration to the Galaxy, although the Antonov is slightly larger. (Paul Osborne)

The long-standing technical, environmental, and cost problems associated with nuclear-powered flight, combined with the problems affecting the basic C-5 program, kept any of the Nuclear Galaxy ideas from becoming a reality. However, even into the mid-1970s there was enthusiasm in some corners for nuclear aircraft, spurred mostly by the oil crisis that had designers looking at alternative fuels. Any thought of actually producing such an aircraft probably died for good after the 1979 nuclear accident at Three Mile Island.

Bomber

By 1967, SAC B-52s armed with conventional "iron" bombs had been engaged in combat in Southeast Asia for several years, with the B-52D models being able to carry dozens of 750-lb bombs. With such conflicts in mind, Lockheed proposed a bomber version of the Galaxy that could outdo even these aircraft, being able to carry no fewer than 288 of these weapons on a dozen pallets. Such an aircraft would have had no chance of survival against even moderate air defenses, and the concept was never pursued.

Although having fewer main wheels than the Galaxy, the Antonov's gear was also designed for poor or semi-prepared airstrips. (Terry Lee/ Aviation Consultants Ltd, UK)

Overall view of the An-124-100 flight deck. (Terry Lee/Aviation Consultants Ltd, UK)

Like the Galaxy, the An-124 dates back before the era of the "glass cockpit" although more modern configurations have been proposed. (Terry Lee/Aviation Consultants Ltd, UK)

The two engineer positions, on the starboard side of the flight deck. (Terry Lee/Aviation Consultants Ltd, UK)

ABCCC & National Airborne Command Post Versions

The conflict in Southeast Asia also spurred the development of the Airborne Command Control and Communications (ABCCC) system for coordinating tactical air to ground operations, the mission being carried out by the C-130E-I/EC-130E. The Galaxy's capacious interior held out the promise of a truly elaborate successor; with three additional fuel tanks fitted in the wings, an "EC-5A" could have stayed aloft for nearly a day without tanker support.

To accommodate the ABCCC systems the fore and aft cargo doors would be sealed; the C-130E-I conversions carried trailer capsules that could be rolled on and off, while the proposed C-5 derivative would have been permanently configured for the role. The extensive communications equipment fitted would have, of course, entailed the installation of many additional aerials to the fuselage and wings.

The USAF never let a formal requirement for a larger ABCCC aircraft, but the Galaxy was also seen as a contender for another command post role. By the late 1960s, the EC-135J aircraft used in the National Emergency Airborne Command Post role were becoming too small for the "Nightwatch" mission, and thought was given to buying a C-5 or 747 as a replacement. The Boeing product ended up being bought as the E-4.

Tanker Versions

Although procurement of the Boeing KC-135 tanker had only ended several years previously, Lockheed was quick to propose the Galaxy as an improved tanker/transport, one that, thanks to its huge size, could do the job of multiple Stratotankers. As originally foreseen, a C-5A tanker version would have used the C-5A wings, powerplants, and tail unit, but a smaller fuselage. Subsequent plans used the standard fuselage, and by 1969 Lockheed ads were showing a Galaxy equipped with a centerline boom and remotely controlled boom pods under the outer wings, refueling a trio of FB-111s simultaneously.

Mothership

At the time the Galaxy was entering service, the USAF was having one of its periodic flirtations with unmanned aerial vehicles, at that time termed RPVs. No small measure of success was being achieved over Southeast Asia by AQM-34s launched from DC-130s, so it was natural that plans for improved systems were being put forth. The Galaxy was considered as a much more capable airborne launcher/control station for drone operations, as was the Boeing 747.

Lockheed ads of the era also mention the possible use of the C-5A as an airborne base for small manned aircraft, essentially acting as a flying aircraft carrier. One "mini fighter" concept is represen-

The radio operator (left) and navigator (right) positions on the port side of the flight deck. (Terry Lee/Aviation Consultants Ltd UK)

RA82047 at a gray Manchester, UK on November 30, 1999. (Terry Lee/Aviation Consultants Ltd, UK)

tative of what could have been carried; a single-engined design with highly-swept wings, this diminutive aircraft would have weighed in at only 4-5 tons, thanks to not having landing gear, heavy armament, or a large fuel load.

Neither the Lockheed nor Boeing plans ever came to fruition, perhaps in part due to the mixed results suffered by previous U.S. parasite aircraft programs. A C-5 or 747 mothership may well have been more successful, given the vast amount of experience gained in aerial refueling operations that present many of the same problems, but this will probably remain a matter of historic conjecture. During the late 1980s it had been suggested by some that C-5s were serving as aerial bases for F-117s and/or other "black" aircraft, but at least in the case of the -117 this was completely untrue.

Other Missile Launcher Concepts

Decades before Ronald Reagan brought the debate over strategic defenses back into public attention with his 1983 "Star Wars" speech, the U.S. military had been struggling with the question of how to best defend the country from the rapidly growing threat of Soviet ballistic missiles.

Although the Army ended up gaining responsibility for the short-lived Spartan/Sprint ABM system at Grand Forks, North Dakota, both the USAF and Navy attempted to get into the picture in the late 1960s, notably with the Seaborne Anti Ballistic Missile Interceptor System (SABIMS), aimed at countering the threat of Soviet Yankee and Delta class missile submarines. Sub-launched missiles, although lacking the accuracy of their land-based coun-

"The Beast awaits its load." RA82047 with its track system deployed for loading. (Terry Lee/Aviation Consultants Ltd, UK)

Rear access - a forty foot truck receiving a twenty-ton machine from the hold of RA82047. (Terry Lee/Aviation Consultants Ltd, UK)

terparts, were capable enough to take out population centers and "soft" military targets, such as SAC airfields, and as they were based far closer to their targets than ICBMs, would give little warning time of an attack.

It was believed by many that in order to detect and intercept SLBMs, especially those fired on depressed trajectories, it would be necessary to use airborne platforms. The C-5's large internal capacity and long endurance would have made it ideal for such a role, although dozens of aircraft would be necessary to maintain continuous patrols over several areas on each coast. Each aircraft would be fitted with a long range radar for detecting SLBM launches and controlling the interceptor missiles, allowing operations independent of vulnerable ground stations. As each Soviet sub would have had at least a dozen missiles, the ABM C-5s would each have been armed with at least an equal number of tube-fired interceptors.

The C-5A was seriously considered as a launch platform for cruise missiles, one of the many potential ALCM carriers considered in the late 1970s following the cancellation of the B-1A bomber. Under consideration was a move away from the idea of a bomber that would penetrate Soviet airspace, relying instead on large transport-type aircraft that could fire massive numbers of ALCMs from standoff ranges.

The Galaxy was prime contender for the role, as around 70 Boeing ALCMs or General Dynamics Tomahawks could be carried within the cargo hold and ejected from the rear ramp. Whereas some competing aircraft would have needed extensive redesign for launch bays and other equipment, the C-5A was well-suited for the mission from the start, and ALCM aircraft could be shifted back to the airlift role if needed. Lockheed actually built some prototype hardware for testing, but as far as is known no ALCM launchings were ever carried out by a C-5.

Similar equipment could have turned the C-141 and even the C-130 into strategic platforms; the Lockheed-California L-1011, McDonnell Douglas DC-10, Boeing 707, and 747 were all proposed for the cruise missile role, although the airliner origins of these types meant that major changes to their airframes would have been necessary.

Ultimately, no move was made to buy any of the "strategic missiler" concepts, with SAC instead moving ahead with the B-1B and plans to put the ALCM on the B-52G/H.

Multi-Role Strategic Aircraft

In 1979, Lockheed made the last of its post C-5A/pre C-5B proposals to restart the Galaxy line. The new design, dubbed the Multi Role Strategic Aircraft (MRSA), was primarily intended for the airlift mission, but the basic airframe would also be adaptable to other roles, including tanker/transport. MRSA would have differed from the C-5 in a number of respects; for example, the fuselage would be shortened to 195 feet in length, although it could be stretched by the installation of two plugs. In place of the Galaxy's TF39 engines, MRSA would have been fitted with uprated CF6 derivatives; the CF6-80A/B versions were available, although the preferred powerplant was the -80C model. Although consuming marginally more fuel than the TF39, the -80C offered a thrust increase of 30%, being rated at 58,000 lbs.

The MRSA was also seen as a potential airborne battleship, carrying long-range AIM-54 Phoenix AAMs to defend SAC strategic bombers or other high-value aircraft. The seemingly implausible notion of using the world's largest military aircraft in the air to air role was not without precedent, as the concept of large "missiler" aircraft had been around for some time; in the early 1960s the USAF had studied arming C-135s with the Eagle missile, a predecessor of the Phoenix. MRSA could also have potentially carried an air-to-air version of the Army's MIM-104 Patriot SAM. This had not been the first time an AAM-armed Galaxy had been considered; several years earlier, as planning for the E-3 AWACS system was going forward, thought was given to using the C-5 or another large aircraft as an adjunct to the E-3 in the continental air defense mission, with the converted transports carrying missiles that would be controlled by patrolling AWACS.

MRSA never flew, although the technically less ambitious C-5N/C-5B program was to appear a few years later.

Other Transport Concepts

Lockheed also investigated more radical transports in the 1970s, driven by the need to provide more fuel economical aircraft that could more successfully compete with ground transportation. One of these was a flying wing spanloader design, able to carry 550,000 lb payloads up to 5,000 miles. Several variants were looked at, including a 1.2 million-pound version powered by six 52,000 lb thrust engines mounted overwing, with auxiliary nozzles to blow the exhaust over the control surfaces for low-speed operations. There would be a central fuselage structure with a visor nose; along with

A forty-ton load being winched out the front, while a cargo of crates in being unloaded out the back. (Terry Lee/Aviation Consultants Ltd, UK)

wing tip doors, this would have been used to load the pressurized cargo spaces within the constant chord, 11 foot-thick wing. At the end of each wing was a T-tail. Thanks to the supercritical wing design, extensive use of lightweight composites, and little fuselage drag, a dramatic increase in fuel efficiency would be achieved.

To get around the problem of providing such an enormous aircraft with conventional runways, Spanloader would have been fitted with air cushion landing gear, permitting operations even from bodies of water if necessary. This would also permit the desired capability of placing men and material as close to the front lines as possible.

The basic Spanloader design would also have been adaptable to other roles, such as an ALBM launcher or a very capable triple-point tanker with the fuel capacity of multiple KC-135As.

Much more conventional were company plans for a trio of high-winged, T-tail transports that would have shared common engines, subsystems, and many airframe components. The basic aircraft, with two CFM56/F108 engines, greatly resembled in size and general configuration the Kawasaki C-1, but would be more powerful than the Japanese aircraft with a 16-ton payload limit. The middle design, aimed at the C-130 market, would have had four engines, a length of 132 feet, and a span of 121 feet. The largest, with a weight of 300,000 lbs, would have been in the C-141 class, and would have had a similar weightlifting capability, although bulkier payloads could be carried thanks to the wider cross-section.

Lockheed was not the only company to investigate new, innovative transport ideas. Boeing looked at laminar flow technology, using wings studded with small openings through which air would be sucked in, greatly reducing drag and boosting fuel economy. Such a configuration was hardly new, as it had been considered for the CX-X of the early 1960s, and Northrop actually flew a pair of X-21s (rebuilt B-66s) with new BLC wings. One Boeing concept had a very "Galaxy-like" fuselage with a high-mounted, high-aspect wing. Mounting engines in underwing pods would be impractical with such an aircraft, as these would conflict with the struts necessary to brace the wing. The solution offered would be to move the engines, of an unspecified type, to the T-tail structure.

The Antonov's built-in crane system lifting a 7-ton crate. (Terry Lee/ Aviation Consultants Ltd, UK.)

Advanced Military Commercial Airlifter

In 1981, even as Lockheed was working on its CX contender and the C-5N, plans were also in the works for an even larger transport, dubbed the Advanced Military Civilian Airlifter. As its name implied, it would carry out heavy lift missions for both the USAF and commercial operators. Such a concept was hardly new, as Lockheed had attempted to sell all three of its airlifters commercially, with only the L-100 version of the Hercules finding sales.

The basic ACMA design was designated LGA-144, and clearly owed much to the Galaxy, using the same basic layout. At 286 feet, the -144 would have been longer than the C-5A and had an increased wingspan of 288 feet. The flight deck was given a 747-style "hump," and the aircraft would ride higher on the ground than the Galaxy. The LGA-144's payload capacity would be boosted to nearly 400,000 lbs.

ACMA's design would be a compromise between military and commercial needs. For example, USAF versions would have had Galaxy type kneeling landing gear for roll-on/roll-off loading, but commercial aircraft could have had this feature deleted for simplicity and weight/cost savings, relying on ramps for loading if carrying out transport duties under contract for the military. ACMA was never built, but ironically the anticipated 1990s heavy lift market did materialize, although it was mostly filled by Ukranian-built Antonov An-124s.

Space Mission Support

As it came into service just as the U.S. space program was reaching its zenith, it is no surprise that the Galaxy has seen its share of use supporting both manned and unmanned missions to orbit and beyond. A C-5 was used to bring in replacement tail fins for the Saturn IB booster that was to launch a Skylab crew, permitting the mission to proceed. However, by this time NASA was already beginning to concentrate on the Space Transportation System, better known in later years as the Space Shuttle.

The logistics of supporting operations by these reusable vehicles would be daunting, as their Saturn IB and Saturn V predecessors, although larger, were able to be broken down into stages for shipment by barge and Super Guppy aircraft, neither method being suitable for transporting the orbiters. Consideration was given to fitting each shuttle with five TF33 turbofans to allow self-deployment capability, but this would have incurred a significant weight penalty and was not done.

Transporting the orbitors on the back of large aircraft was seemingly the only other reasonable method of quickly moving the shuttles between the assembly/refurbishment facility at Palmdale, California, the Cape, and the landing site at Edwards AFB.

Permanently modifying the C-5A for shuttle transport would have started with cutting holes in the top of the fuselage, and in the wing root fillings, at stations 665 and 1605, respectively. This would allow trusswork to be fitted in the cargo compartment, redistributing the weight of the orbiter to the floor and upper compartment walls of the Galaxy. The orbiter attachment structure would be fitted at these points prior to a ferry mission, adding just under a ton to the aircraft's weight. Aside from the ferry flights of operational orbiters, there was also the necessity to carry out atmospheric flight tests to verify that the shuttle could make unpowered landings after its return from orbit.

There were also more radical shuttle/C-5 systems proposed, such as extensively modifying the transport's fuselage to carry an orbiter semi-internally, or towing the spacecraft behind a Galaxy.

Also investigated was the possibility of carrying the shuttle's large external fuel tank on either a C-5 or 747. Of similar dimensions to the orbiter vehicle, the tank had a dry weight of 78,000 lbs, although fitting a fairing over the tank's blunt bottom and installing other ferry hardware would raise this figure by six tons. As it turned out, air transport of the tanks never went forward, and they have always been brought to the launch site by barge.

There was also some interest in building a purpose-designed aircraft for STS transport and drop tests, since it was not certain early on in the program that either the C-5 or 747 would be able to safely drop an orbiter for the approach and landing trials. John Conroy, who had built the Guppy/Pregnant Guppy/Super Guppy outsized transports proposed an even more radical aircraft for this task. Called Virtus, this 450-foot span monster would have carried an orbiter under a high-mounted wing and between two fuselage pods, with the port structure mounting a surplus C-97 cockpit and forward fuselage. Each pod would be fitted with B-52 landing gear, giving a crosswind capability.

A small part of the payload - the larger crates are seven tons each, while the smaller ones weigh in at four tons. (Terry Lee/Aviation Consultants Ltd, UK.)

Early studies showed the Virtus to be powered by four Starlifter-type TF33-P-7 turbofans; much more powerful JT9D engines were later substituted. There were also studies for a derived version, called Collosus, using the same wing, tail unit, and C-97 cockpit, but with a single massive fuselage for outsized payloads of up to half a million pounds. Collosus would have a gross weight of up to 1.2 million pounds, a figure not achieved in reality until the arrival of the Antonov An-225 nearly two decades later.

In May 1975, NASA announced that it had chosen the 747 for the shuttle carrier mission; although both aircraft had been deemed capable of carrying out the role, the space agency was concerned that the USAF could not guarantee availability of modified C-5As.

Although the Galaxy would never carry the shuttle, C-5s would continue to provide major support for NASA efforts, as well as spawn several unbuilt space mission derivatives. Several years later, one mission proposed for the MRSA derivative of the C-5A was an airborne space vehicle launcher, using either Minuteman-derived expendable boosters dropped from the cargo hold, or manned "mini shuttles" carried dorsally. The Galaxy has also been considered as one of the possible candidates to replace NASA's long-serving NB-52B launch aircraft, although this requirement is a long-term one.

C-5C Space Cargo Modified

Quite aside from the logistical demands of transporting elements of the space shuttle system across the country, from early on in the program it was recognized that getting the shuttle's payloads to the Kennedy Space Center also posed problems, as it was planned to launch large items that would not fit internally on either a C-5 or 747. Road or sea transport of these extremely sensitive (as well as expensive) items was not always practical. To meet this need, Boeing in 1974 proposed a 747 modified to carry a large dorsal cargo pod, which could be used to carry shuttle payloads. Although the 747 did end up getting the shuttle carrier mission, the cargo carrier plan never went through.

The need for air transportation still remained, however, and in the late 1980s a pair of C-5As were taken aside for the Space Cargo Modification program. Redesignated as C-5Cs, the SCM aircraft (nicknamed "Scums") are configured to carry a special trailer that approximates the space shuttle's cargo bay. At 17 ft in height, this trailer is too tall to fit in a standard C-5A, so the SCM aircraft have had their rear upper cargo compartments removed, a major structural revision that also entailed modifying and moving to the aft the rear pressure bulkhead and beefing up a longeron. The aft cargo door also had to be modified, with the door now being split down the center and attached to the petal doors.

A major role for the retrofitted aircraft was to be the transportation of space station components; although some of these could be transported by road, many could not be, and shipment by sea was too time-consuming. NASA's small fleet of Super Guppy aircraft had the volume capacity for these loads, but were rapidly approaching the end of their service lives, with some of their airframe components dating back to the 1950s.

The SCM aircraft were redelivered to the USAF following modification by Lockheed in 1988-89; initial user unit was the 433rd, but they were later reassigned to the 60th at Travis. There were plans to convert at least a half-dozen additional C-5As to SCM configuration to support the airlift of large special operations helicopters, such as the MH-47 and MH-53, without disassembly, but this has not been carried out to date.

Not all spacecraft lift missions have been within the United States; for example in the spring of 1985, a C-5A flew to Bremen, Germany, to pick up the European Spacelab D-1 module for transport to KSC, with support equipment being brought over by a chartered 747. And on August 22, 1997, the NASA-Japanese Tropical Rainforest Measurement Mission, built in the U.S. was shipped aboard a Galaxy from Andrews AFB, Maryland, to Kagoshima, Japan. In order to minimize vibration on the delicate payload, the mission was made nonstop with multiple air refuelings.

Lockheed Martin itself has leased C-5s from the USAF to take Atlas III boosters from the company's San Diego plant to the former Martin Marietta facility near Denver, as well as completed rockets from Colorado to Florida.

Other spacelift missions by C-5s have included:

Hardware for STS-103, the third shuttle mission to repair the Hubble Space Telescope. A decade earlier, the Hubble itself had been airlifted by a Glaxy from Palmdale, California, to Florida.

Atlas-I booster for the GOES-K weather satellite.

Node 1 module for the International Space Station.

Centaur upper stage for the Cassini probe.

The Chandra X-Ray observatory satellite.

Atlas-IIA rocket and Centaur upper stage for the GOES-L satellite.

Megalifter

The tale of Galaxy derivatives would not be complete without the inclusion of the Megalifter, a 1974 concept that, had it been built, would have dwarfed the C-5 as well as every other heavier-than-air craft ever built. Presaging Lockheed Skunk Works efforts of a quarter-century later, the Megalifter was to be a semi-heavier than air design, incorporating the features of both airships and conventional aircraft to carry extremely outsized and heavy loads.

The lifting body fuselage would be of geodetic construction, holding cells for seven million cubic feet of helium. the gas would generate part of the lift at takeoff, with the fuselage and wings providing the rest necessary for flight. Megalifter's dimensions were to be staggering; total length was 650 feet, or about the same as a WWII-era aircraft carrier. Wingspan was over 500 feet, or equivalent to three B-52s stood wingtip to wingtip. Without any gas or payload, the Megalifter would weigh in at 725,000 lbs, but the helium would have an equivalent buoyancy of some 239 tons, giving

The An-124's 51,000lb thrust D18T turbofans are significantly more powerful than the Galaxy's TF39s, but there have been proposals to reengine the Antonov with more reliable and fuel-efficient engines of western origin. (Paul Osborne)

a filled aircraft an equivalent weight of nearly a quarter-million pounds.

Megalifter's keel would be formed by an immense cargo compartment some 40 feet high, 40 feet wide, and 300 feet in length. The floor of the cargo compartment would consist of removable pallets that could be fitted as necessary; aircraft up to the size of a space shuttle orbiter could be accommodated in a semi-submerged fashion. Total payload would be around 400,000 lbs, an impressive figure even without considering that it would be equivalent to twice the aircraft's effective unloaded weight. The aircraft could also have served as a very long duration airborne ICBM launcher, command post, or relay station.

Although not a Lockheed project, Megalifter, at least in prototype form, could have made maximum use of components stripped from the damaged C-5A 67-0712 airframe. These included the forward fuselage and the cockpit able to be readily converted for use, while the visor nose would allow for easy access to the cargo compartment.

One powerplant option for Megalifter would have been four TF39s, coupled with a pair of smaller outboard engines for thrust vectoring. Megalifter's 200 mph speed, while a fraction of that of the Galaxy, would nonetheless allow it to reach Europe within a single day, far faster than fast cargo ships. The semi-LTA configuration also offered a range of 10,000 miles, sufficient to reach virtually any trouble spot in the world nonstop and without tanker support. Despite Megalifter's immense proportions, airfield operations were to be eased by the aircraft's STOL capabilities and low takeoff/landing weight.

The Megalifter was never to be built, despite the increase in airlift capacity that it offered. However, the soundness of the basic idea can be judged by the disclosure over two decades later of a similar Lockheed Skunk Works project, which will be discussed later in this chapter.

Advanced Mobility Aircraft
Looking beyond the C-5M retrofit program, by the late 1990s Lockheed Martin was also engaged in the Advanced Mobility Aircraft project, an in-house study aimed at providing replacements for several types of large military aircraft in the 21st Century.

Lockheed first looked at designs not dissimilar to the Galaxy and Starlifter, with such features as a 25-swept wing with four engines and a T-tail. Also considered were twin-engine aircraft with supercritical wings that had a reduced degree of sweep.

Such designs were considered for a number of roles, including a strategic airlifter to replace C-141 class aircraft, and a "KC-X" tanker-transport to replace the KC-135R and converted 707s. Both aircraft would have shared much of the same structure, with the tanker being equipped with a Flying Boom and underwing drogue pods.

Box Wing
Moving away from designs with conventional airlifter planforms, Lockheed Martin also investigated more radical concepts, such as a tanker/transport fuselage married to a "box wing," this consisting of a swept-back forward section fitted low on the forward fuselage, joined at the tip by a high-mounted, swept-forward rear unit.

Blended Body Transport
Another radical concept for AMA was a transport with a blended wing/box structure, with the engines being mounted atop the fuselage. As was the case with the C-5A decades before, Lockheed is proposing derivatives of the basic design for other missions. Significantly, the basic planform used is amenable to being refined to a low-observable configuration, which would make it a contender for the special operations tanker/transport role, as the USAF's Special Operations Command has identified the need for an "MC-X" aircraft to replace its aging MC-130 *Combat Talon I* and *Combat Shadow* fleets. The design could also be adapted for the long-range strike role carrying standoff precision-guided missiles; a new-generation of Unmanned Aerial Vehicles could also be carried, with the transport serving as a mothership for launch and control of the UCAVs.

Aerocraft
Resurrecting the idea of a semi-lighter than air ultra-large airlifter, Lockheed Martin announced in 1999 that its Skunk Works division was working on just such a design, termed the Aerocraft. Like the Megalifter of two decades before, Aerocraft would derive a portion of its lift (around 50%) from helium, with the remainder coming from a lifting body fuselage nearly 800 feet in length. Four engines of an unspecified type would be fitted to the fuselage in a tilt-rotor configuration to permit STOL performance, while top speed would be around 125 kts. The extremely wide fuselage would accommodate four frontal door/ramps, and total payload would be at least a million pounds.

Galaxy Counterparts
For more than a decade after its first flight, the Galaxy hung onto its rank as the world's largest military airlifter; there was no development of a rival aircraft in the western bloc, but behind the "Iron Curtain" it was a different story. Russia had had a long history with the design of large transport aircraft; indeed, the Sikorsky *Russki Vitiaz/Le Grand* of 1913 was the world's first four-engined aircraft. After the revolution, the fascination with giant aircraft continued, as exemplified by the eight-engined Maxim Gorky eight-engined transport, which had a wingspan only sixteen feet less than that of the C-5A.

With this heritage it is no surprise that the Soviets (and later the Russians and Ukranians) have produced designs to rival, and even exceed, the Galaxy. Although there was, of course, a need for transporting heavy military loads, this was not the only concern. With men, material, and a road/rail network already in place in Eastern Europe, there was no need for rapid air deployment across intercontinental distances. However, unlike the U.S. with its vast domestic road and rail network, the USSR had immense reaches of territory unreachable by any means but air.

The sole An-225, taking off from RAF Farnborough. The world's largest flying aircraft for over a decade, the Mriya has been grounded, and is said to be serving as a parts repository for An-124 operations. (Nick Challoner)

Although studies for what became the An-22 began only a year or so before the C-X program started, the Soviet aircraft was actually in many ways more a contemporary with the canceled XC-132 of a half-decade before. Rather than create an entirely new design, Antonov, like Lockheed, made maximum use of previous airlifter design experience, dramatically scaling up the An-12 Cub airframe and adding a double tail unit. Whereas the Cub had a straight wing, the An-22 had a moderately-swept wing; the powerplants were four colossal Kuznetsov NK12 turboprops, originally developed in the early 1950s for the Bear bomber, and each rated at 15,000 shp.

There were, of course, some features that the American and Soviet aircraft had in common; as the An-22 would also have to operate from all manner of surfaces, the Antei was also fitted with high-flotation gear, in this case three twin-wheel main gears fitted in tandem. Like other Soviet transport designs of the time, the An-22 has a glazed "bombardier" type nose; a radome just aft of the glazing was later fitted.

The first An-22 flight took place on February 27, 1965, more than three years before the Galaxy first took to the air.

Like the Galaxy, the An-22 was considered as an ultra-large airliner, in this case capable of carrying 700 passengers in a double-deck configuration that was not proceeded with. However, many did fly in civil Aeroflot colors, although these were, of course, available for military duty.

Military equipment able to be hauled aboard the An-22 included such items as ASU-85 assault guns, T-65 main battle tanks, and mobile SAM batteries. An-22s took supplies to Arab client countries during the tense years of the early 1970s, as well as transporting armaments to the government in Angola during the fighting there. The USSR's own military operations relied on the An-22, including the 1968 invasion of Czechloslavakia and the move 11 years later in Afghanistan.

Although some An-22s were still at least nominally in service by the late 1990s, development by Antonov of a successor design had begun more than two decades previously. First flown on December 26, 1982, but not shown in the West until the Paris Air Show of 1986, the An-124 *Condor* is easily recognizable as having drawn much of its inspiration from the Galaxy, differing in general layout only by the fitting of a low-mounted tailplane.

Internal arrangements are basically similar to those of the American aircraft, with a relief crew area aft of the flight deck, and an 88-passenger upper compartment behind the wing. There are differences, however, including the cargo compartment that is pressurized to a lower differential than the upper compartments, unlike the Galaxy, whose hold is fully pressurized. The Antonov does not typically carry personnel in the hold, but during the run-up to the Persian Gulf War in 1990 one aircraft did so, carrying 451 passengers on an evacuation flight out of Amman, Jordan.

The *Condor* is powered by four Ivchenko Progress/Lotarev D18T turbofans; rated at 51,950 lbs of thrust each, these are significantly more powerful than the Galaxy's TF39s. Unlike its American counterpart, the *Condor* has no provision for taking on fuel while airborne, although its range on internal fuel is far-spanning, with a maximum ferry distance of over 8,000 nautical miles. Even longer flights have been made; in 1987 an An-124 made a nonstop unrefueled flight around the borders of what was then the USSR, covering some 12,000 miles. The design's weightlifting capabilities were demonstrated on July 1985 when an early An-124 took a 377,463 lb payload to an altitude of 35,269 feet.

The An-124 has achieved no small degree of commercial success, with several operators flying the type worldwide, on such diverse missions as supporting a world tour by singer Michael Jackson, to ferrying disassembled Nimrod MR.2s (themselves no mean aircraft) to Bournemouth, and later Woodford for rebuilding into the MR.4 configuration.

The popularity of the An-124 in commercial service led to plans for "westernized" derivatives; the An-124-100M was to have some Western avionics, while the An-124-102 is a projected model with a three-man glass cockpit.

Even western air arms have made use of chartered An-124s. The Royal Air Force, with no outsized airlift capability of its own (despite long-standing plans to buy or lease C-17s) used contracted -124s to ferry equipment to Tornado squadrons operating in Saudi Arabia, and in late 1999 it was reported that an An-124 had been used to transport British Chinook helicopters to Oman and back. Earlier that year, three Dutch Air Force Chinooks had been brought home from duty in Albania aboard a *Condor*. An-124 have also taken Tornado F.3s to and from duty in the Falklands. Procurement of the An-124-210 version was a proposed solution for the RAF's Short-Term Strategic Airlift requirement, this derivative to have been fitted with new avionics and British Rolls-Royce RB.211-524EH or -524HT engines. The STSA program was canceled in the late summer of 1999, although the need for such an aircraft re-

The Mriya's primary role was to be the carriage of heavy loads such as the Buran space shuttle externally, hence the new twin tail. (Nick Challoner)

mained. On May 16, 2000, the RAF announced that it would lease a quartet of Boeing C-17s for the outsized airlift mission.

Aircraft and armaments of non-western origin have also been carried by An-124s, including delivery flights of Su-27s to Vietnam, MiG-29s to Malaysia, and Su-30s and T-90 tanks to India.

Like the Galaxy, *Condors* have been used to expedite the transportation of spacecraft and their systems. For example, on June 7, 1995, an An-124 arrived at the Kennedy Space Center carrying a docking module to be fitted to the space shuttle *Atlantis* for the STS-74 docking with the Russian Mir space station. Arianespace has also contracted An-124 flights, such as the movement of the Aerospatiale Sirius 2 satellite from France to the Ariane launch center in French Guyana.

Whereas Lockheed's plans for the larger ACMA derivative of the C-5A did not proceed, Antonov actually did build an enlarged version of the -124. Rolled out on November 30, 1988, the An-225 *Cossack* easily claimed the title of the world's largest aircraft, with a wingspan of 290 feet and a length of over 275 feet.

Although clearly a *Condor* derivative, the -225 is even more impressive, with a new, wider wing center section to permit the mounting of an additional pair of D18T engines. The fuselage is some 23 feet longer than that of the An-124, and as the An-225 was designed with the primary role of carrying external cargo, Antonov took the opportunity to fit an entirely new double tail, which is less affected by airflow disturbances from dorsal loads. As with some of the unbuilt commercial Galaxy derivatives, there is no rear door, the roll-on/off capability not being needed. Maximum takeoff weight is a staggering 1,322,770 lb; to cope with this an additional pair of main wheels are fitted.

The main purpose behind the creation of such a huge design was support of the Buran space shuttle and the associated Energia heavy-lift space launcher. There had been several conversions of Mysasichev *Bison* bombers for carrying either a *Buran* or the oxygen or hydrogen tank components of the *Energia*, but a more capable purpose-built aircraft to replace these was desired.

Originally, several An-225s were to be built, but the cancellation of both the *Buran* and *Energia* programs left the type without a primary mission. By 1995, construction was underway on a second example, but this was never completed. The An-225 was touted as an airborne launcher for other spacecraft, including the BAe Hotol shuttle and the Svityaz rocket, but to date no such operations have taken place. Plans for other types of use may never come about, as in 1999 the sole aircraft was reported to have been taken out of service.

Although using the same basic D-18T turbofans as its An-124, the Mriya had more of them, with a new center wing section allowing two additional engines to be installed to handle the aircraft's gross weight of over one million pounds. (Nick Challoner)

Among the many as-yet-unbuilt aircraft projects that appeared from designers in the former USSR during the 1990s were several even larger transports, ironically using the basic configuration of the Virtus aircraft proposed more than two decades previously. The NPO Molniya group proposed the Molniya 100 Heracles, with cockpits on both fuselage booms and a very large canard surface. Powered by six 50,000 lb thrust-class engines, the Heracles would weigh just under two million pounds on takeoff, while carrying a 990,000 lb payload. Myasishchev planned the M90 Air Ferry, a propfan aircraft with a portside cockpit. Available in six- or eight-engined models, the Air Ferry in its largest form would carry 880,000 lbs, with a gross weight of 1.8 million pounds.

Appendix

U.S. Strategic Airlifter Serial Number Dispositions

C-124 Globemaster II
48-0795
50-0083/0118 (C-124A)
50-1255/1268 (C-124A)
51-0072 (YC-124B)
51-0073/0182 (C-124A)
51-5173/5187 (C-124A)
51-5188/5213 (C-124C)
51-7272/7285 (C-124C)
52-0939/1089 (C-124C)
53-0001/0052 (C-124C)

Preserved: 51-0089, 52-0943, 52-0994, 52-1000, 52-1066, 52-1072, 53-0044, 53-0050

C-133 Cargomaster
54-0135/0146 (C-133A)
56-1998/2014 (C-133A)
57-1610/1615 (C-133A)
59-0522/0536 (C-133B)

Attrition: 54-0140, 54-0146, 56-2002, 56-2005, 56-2014, 57-1611, 57-1614, 59-0523, 59-0530, 59-0534

Preserved: 56-2008, 56-2009, 59-0527

Starlifter Serial Numbers
61-2275/2779
63-8075/8090
64-0609/0653
65-0216/0281
65-9397/9414
66-0126/0209
66-7944/7959
67-0001/0031
67-0164/0166

61-2779 wore the tail band of the new Air Force Materiel Command in June 1992. (Chris Reed)

NC-141A *Against the Wind*. (Chris Reed)

87-0027 of the 436 AW, May 1997 (Robbie Robinson)

Known dispositions

Serial	Type	Disposition	Date	Status
61-2775	(JC-141A) (NC-141A)	Preserved AMC Museum, Dover AFB		
61-2776	(NC-141A)	Arrived at AMARC	7-AUG-1998	Still there 3-JAN-2000
61-2777	(NC-141A)	Arrived at AMARC	27-SEPT-1994	Still there 3-JAN-2000
61-2779	(NC-141A)	Preserved Edwards AFB		
63-8075	(C-141B)	Preserved Travis AFB		
63-8078	(C-141B)	Arrived at AMARC	12-DEC-1995	Still there 29-OCT-1999
63-8079	(C-141B)	Preserved Charleston AFB		
63-8082	(C-141B)	62 AW	JAN-2000	
63-8083	(C-141B)	Arrived at AMARC	10-APR-1996	Still there 29-OCT-1999
63-8086	(C-141B)	Arrived at AMARC	19-DEC-1997	Still there 29-OCT-1999
63-8087	(C-141B)	Arrived at AMARC	23-APR-1999	Still there 29-OCT-1999
63-8089	(C-141B)	Arrived at AMARC	22-OCT-1996	Still there 29-OCT-1999
63-8090	(C-141B)	Arrived at AMARC	07-AUG-1996	Still there 29-OCT-1999
64-0609	(C-141B)	Arrived at AMARC	23-JUL-1997	Still there 29-OCT-1999
64-0610	(C-141B)	Arrived at AMARC	15-MAR-1999	Still there 29-OCT-1999
64-0613	(C-141B)	Arrived at AMARC	24-NOV-1997	Still there 29-OCT-1999
64-0617	(C-141B)	Arrived at AMARC	15-MAR-1996	Still there 29-OCT-1999
64-0619	(C-141B)	305 AMW	JUNE-1999	
64-0623	(C-141B)	Arrived at AMARC	14-SEP-1999	Still there 29-OCT-1999
64-0625	(C-141B)	Arrived at AMARC	22-OCT-1997	Still there 29-OCT-1999
64-0632	(C-141B)	172 AW/MS ANG	JAN-2000	
64-0634	(C-141B)	Arrived at AMARC	23-APR-1996	Still there 29-OCT-1999
64-0635	(C-141B)	Arrived at AMARC	16-JUN-1997	Still there 29-OCT-1999
64-0636	(C-141B)	Arrived at AMARC	08-JUN-1993	Still there 29-OCT-1999
64-0638	(C-141B)	305 AMW	JUNE-1999	

(Tim Doherty)

(Tim Doherty)

(Tim Doherty)

(Tim Doherty)

64-0639	(C-141B)	Arrived at AMARC	20-JUN-1997	Still there 29-OCT-1999
64-0640	(C-141B)	(C-141C) 183 AS/MS ANG	MAY-1999	
64-0644	(C-141B)	305 AMW	MAY-98	
64-0645	(C-141B)	756 AS/AFRC	JUNE-1999	
64-0646	(C-141B)	305 AMW	MAY-98	
64-0648	(C-141B)	Arrived at AMARC	01-JUL-1993	Still there 29-OCT-1999
64-0649	(C-141B)	305 AMW MAY-98	437 AW JAN-2000	
64-0650	(C-141B)	Arrived at AMARC	03-OCT-1995	Still there 29-OCT-1999
64-0651	(C-141B)	Arrived at AMARC	21-NOV-1996	Still there 29-OCT-1999
65-0217	(C-141B)	305 AMW	MAY-98	
65-0223	(C-141B)	Arrived at AMARC	01-OCT-1998	Still there 29-OCT-1999
65-0226	(C-141B)	756AS/AFRC	JUNE-1999	
65-0227	(C-141B)	Arrived at AMARC	18-SEP-1998	Still there 29-OCT-1999
65-0234	(C-141B)	Arrived at AMARC	02-APR-1999	Still there 29-OCT-1999
65-0237	(C-141B)	89 AS/445 AW	JULY-98	
65-0242	(C-141B)	McGuire SEPT-98	Arrived at AMARC 05-APR-1999	Still there 29-OCT-1999
65-0243	(C-141B)	Arrived at AMARC	06-APR-1999	Still there 29-OCT-1999
65-0247	(C-141B)	Arrived at AMARC	15-SEP-1995	Still there 29-OCT-1999
65-0252	(C-141B)	Arrived at AMARC	21-DEC-1998	Still there 29-OCT-1999
65-0256	(C-141B)	(C-141C)		
65-0260	(C-141B)	Arrived at AMARC	05-AUG-1999	Still there 29-OCT-1999
65-0261	(C-141B)	(C-141C)	445 AW/AFRC JAN-2000	
65-0262	(C-141B)	Arrived at AMARC	13-JUL-1993	Still there 29-OCT-1999
65-0263	(C-141B)	Arrived at AMARC	15-JAN-1999	Still there 29-OCT-1999
65-0264	(C-141B)	Arrived at AMARC	18-MAR-1996	Still there 29-OCT-1999
65-0265	(C-141B)	Arrived at AMARC	24-APR-1996	Still there 29-OCT-1999

(Arno J.A.H. Cornelissen)

(Arno J.A.H. Cornelissen)

C-5B -40061 of the 436th AW.

C-5A -90004 of the 436th MAW.

65-0266	(C-141B)	305 AMW	JUNE-1999	
65-0268	(C-141B)	Arrived at AMARC	21-APR-1997	Still there 29-OCT-1999
65-0269	(C-141B)	437 AW	DEC-99	
65-0270	(C-141B)	Arrived at AMARC	19-NOV-1997	Still there 29-OCT-1999
65-0271	(C-141B)	756 AS/AFRC	JUNE-1999	
65-0272	(C-141B)	Arrived at AMARC	08-JUN-1998	Still there 29-OCT-1999
65-0273	(C-141B)	305 AMW	JUNE-1999	
65-0275	(C-141B)	Arrived at AMARC	17-DEC-1998	Still there 29-OCT-1999
65-0278	(C-141B)	Arrived at AMARC	05-JAN-1996	Still there 29-OCT-1999
65-0280	(C-141B)	305 AMW	JUNE-1999	
65-9397	(C-141B)	Arrived at AMARC	03-APR-1996	Still there 29-OCT-1999
65-9398	(C-141B)	Arrived at AMARC	20-MAY-1993	Still there 29-OCT-1999
65-9399	(C-141B)	Arrived at AMARC	10-OCT-1995	Still there 29-OCT-1999
65-9402	(C-141B)	Arrived at AMARC	16-JAN-1996	Still there 29-OCT-1999
65-9403	(C-141B)	62nd AW		
65-9404	(C-141B)	Arrived at AMARC	09-DEC-1997	Still there 29-OCT-1999
65-9409	(C-141B)	89 AS/445 AW	JULY-1998	
65-9411	(C-141B)	305 AMW	MAY-1998	
65-9412	(C-141B)	89 AS/445 AW	JULY-1998	
65-9413	(C-141B)	305 AMW	JUNE-1999	
66-0128	(C-141B)	Arrived at AMARC	10-SEP-1998	Still there 29-OCT-1999
66-0129	(C-141B)	Arrived at AMARC	17-APR-1996	Still there 29-OCT-1999
66-0131	(C-141B)	305 AMW	JUNE-1999	
66-0133	(C-141B)	89 AS/445 AW	JULY-98	
66-0135	(C-141B)	Arrived at AMARC	22-DEC-1997	Still there 29-OCT-1999
66-0137	(C-141B)	62 AW	MAY-98	
66-0138	(C-141B)	Arrived at AMARC	11-JAN-1996	Still there 29-OCT-1999

C-5B -400060 of the 60th AW.

C-5B -50006 of the 60th AW.

66-0141	(C-141B)	Arrived at AMARC	04-DEC-1997	Still there 29-OCT-1999
66-0143	(C-141B)	Arrived at AMARC	11-MAY-1993	Still there 29-OCT-1999
66-0145	(C-141B)	Arrived at AMARC	23-DEC-1996	Still there 29-OCT-1999
66-0153	(C-141B)	756 AS/AFRC	JUNE-1999	
66-0155	(C-141B)	(SOLL-II) 437 AW	JAN-2000	
66-0159	(C-141B)	97 AMW	MAY-1999	
66-0162	(C-141B)	305 AMW	JUNE-1999	
66-0163	(C-141B)	305 AMW	JUNE-1999	
66-0166	(C-141B)	305 AMW	JAN-2000	
66-0168	(C-141B)	305 AMW	JAN-2000	
66-0170	(C-141B)	Arrived at AMARC	08-JUL-1993	Still there 29-OCT-1999
66-0171	(C-141B)	62 AW	JUNE-1999	
66-0172	(C-141B)	Arrived at AMARC	14-OCT-1998	Still there 29-OCT-1999
66-0177	(C-141B)	89 AS/445 AW	JULY-1998	*Hanoi Taxi*
66-0178	(C-141B)	305 AMW MAY-1998	Arrived at AMARC 16-DEC-1998	Still there 29-OCT-1999
66-0179	(C-141B)	Arrived at AMARC	29-JUL-1997	Still there 29-OCT-1999
66-0180	(C-141B)			Preserved Robins AFB
66-0185	(C-141B) (C-141C)	172 AW/MS ANG	JAN-2000	
66-0188	(C-141B)	Arrived at AMARC	22-JUL-1993	Still there 29-OCT-1999
66-0194	(C-141B)	305 AMW	MAY-98	
66-0198	(C-141B)	Arrived at AMARC	18-OCT-1999	Still there 29-OCT-1999
66-0200	(C-141B)	Arrived at AMARC	15-APR-1998	Still there 29-OCT-1999
66-0202	(C-141B)	305/514 AMW	JAN-2000	
66-0203	(C-141B)	Arrived at AMARC	30-DEC-1997	Still there 29-OCT-1999
66-0204	(C-141B)	Arrived at AMARC	25-NOV-1997	Still there 29-OCT-1999
66-0205	(C-141B)	Arrived at AMARC	03-JAN-1997	Still there 29-OCT-1999
66-0207	(C-141B)	Arrived at AMARC	01-OCT-1997	Still there 29-OCT-1999
66-0208	(C-141B)	Arrived at AMARC	28-MAR-1997	Still there 29-OCT-1999
66-0232	(C-141B)	89 AS/445 AW	JULY-98	
66-0257	(C-141B)	89 AS/445 AW	JULY-98	
66-7945	(C-141B)	Arrived at AMARC	18-APR-1996	Still there 29-OCT-1999
66-7946	(C-141B)	Arrived at AMARC	20-JUL-1999	Still there 29-OCT-1999
66-7949	(C-141B)	Arrived at AMARC	28-APR-1999	Still there 29-OCT-1999
66-7956	(C-141B)	62 AW	JUNE-1999	
66-7958	(C-141B)	Arrived at AMARC	31-MAR-1998	Still there 29-OCT-1999
67-0001	(C-141B)	Arrived at AMARC	23-DEC-1999	
67-0002	(C-141B)	62 AW	MAY-98	
67-0005	(C-141B)	Arrived at AMARC	21-NOV-1997	Still there 29-OCT-1999

C-5B -80225 at Westover.

67-0007	(C-141B)	Arrived at AMARC	29-SEP-1999	Still there 29-OCT-1999
67-0009	(C-141B)	Arrived at AMARC	29-JUL-1998	Still there 29-OCT-1999
67-0010	(C-141B)	305 AMW	JUNE-1999	
67-0014	(C-141B)	305 AMW	JUNE-1999	
67-0019	(C-141B)	305 AMW	MAY-1998	
67-0023	(C-141B)	Arrived at AMARC	06-OCT-1995	Still there 29-OCT-1999
67-0025	(C-141B)	Arrived at AMARC	11-APR-1997	Still there 29-OCT-1999
67-0026	(C-141B)	Arrived at AMARC	19-AUG-1999	Still there 29-OCT-1999
67-0031	(C-141B)	89 AS/445 AW	JULY-1998	
67-0164	(C-141B)	Arrived at AMARC	06-NOV-1998	Still there 29-OCT-1999
67-0166	(C-141B)	305 AMW	JUNE-1999	
67-6947	(C-141B)	305 AMW	JUNE-1999	

Attrition: 63-8077, 64-0641, 65-0274, 65-0281, 66-0127, 67-0006, 67-0008

Galaxy Serial Numbers
66-8303/8307 (C-5A)
67-0167/0174 (C-5A)
68-0211/0228 (C-5A)
69-0001/0027 (C-5A)
70-0445/0467 (C-5A)
70-0468 (C-5A, contract cancelled)
71-0180/0212 (C-5A, contract cancelled)
72-0099/0112 (C-5A, contract cancelled)

83-1285 (C-5B)
84-0059/0062 (C-5B)
85-0001/0010 (C-5B)
86-0011/0026 (C-5B)
87-0027/0045 (C-5B)

Known Dispostions:

C-5A

66-8303	w/o OCT-1970	
66-8304	337 AS	NOV-98
67-0167	337 AS	NOV-98
67-0172	c/n 11	w/o MAY 26-1970
67-0173	Stewart	NOV-1998
68-0215	337 AS	NOV-1998
68-0218	w/o APRIL 4-1975	
68-0221	68 AS	DEC-1999
68-0222	337 AS	NOV-1998
68-0227	w/o SEPT-74	
69-0002	433 AW	OCT-1998
69-0003	337 AS	NOV-1998
69-0004	433 AW	OCT-1998
69-0005	337 AS	NOV-1998
69-0006	433 AW	JULY-1998
69-0011	337 AS	NOV-1998
69-0012	137 AS	JULY-1998
69-0015	Stewart	APRIL-1999
69-0017	337 AS	NOV-1998
69-0020	337 AS	NOV-1998
69-0022	337 AS	NOV-1998
69-0026	60 AMW	MAY-1999
70-0447	436 AW	JUNE-1999
70-0448	337 AS	NOV-1998
70-0452	97 AMW	MAY-1999
70-0453	97 AMW	MAY-1998
70-0456	60 AMW	MAY-1998
70-0457	436 AW	OCT-1998
70-0459	60 AMW	MAY-1998
70-0461	436 AW	JUNE-1999
70-0462	60 AMW	MAY 1998
70-0464	60 AMW	MAY-1998

C-5B

83-1285	436 AW	JUNE-1999
84-0060	60 AMW	OCT-1998
84-0062	60 AMW	OCT-1998
85-0001	436 AW	DEC-1999
85-0003	436 AW	DEC-1999
85-0004	60 AMW	OCT-1998
85-0005	436 AW	JUNE-1999
85-0006	60 AW	DEC-1999
85-0007	436 AW	JAN-2000
85-0010	60 AW	
86-0002	436 AW	JUNE-1999
86-0011	436/512 AW	OCT-1998
86-0012	60 AMW	MAY-99
86-0013	436 AW	DEC-1999
86-0014	60 AMW	OCT-1998
86-0016		
86-0017	436 AW	JAN-2000
86-0018	60 AMW	MAY-1999
86-0019	436 AW	DEC-1999
86-0020	436 AW	DEC-1999
86-0022	60 AMW	OCT-1998
86-0023	436 AW	JUNE-1999
86-0024	60 AMW	JAN-1999
86-0025	436 AW	JAN-2000
87-0027	436 AW	DEC-1999
87-0028	60 AMW	OCT-1998
87-0029	436 AW	JAN-2000
87-0030	60 AMW	JAN-2000
87-0031	436 AW	JAN-2000
87-0032	60 AMW	JUNE-1999
87-0033	436 AW	JAN-2000
87-0034	436 AW	MARCH-1999
87-0035	436 AW	JUNE-1999
87-0036	60 AMW	JAN-2000
87-0043	436 AW	JAN-2000
87-0044	60 AMW	OCT-1998
87-0045	436/512 AW	JAN-2000

C-17 Globemaster III Serial Numbers
87-0025
88-0265/0266
89-1189/1192
90-0532/0535
92-3291/3294
93-0599/0604
95-0102/0107
96-0001/0008
97-0041/0048
98-0049/0057

99-0058/0070
02-0101/0115
03-0116/0120

Known dispositions:

87-0025	**"T-1"**		
87-0048	**437 AW**	**JAN-2000**	
88-0265	**"P-1"**	**Payload to altitude records**	
88-0265	**97 AMW**	**MARCH-1999**	
89-1189	**437th AW**		
89-1190	**437th AW**		
89-1191	**437th AW**		
89-1192	**First a/c to AMC**		
90-0533	**437th AW**		
90-0533	**97 AMW**	**MARCH-1999**	
92-3291	**First operational overseas flight**		
92-3293	**97 AMW**	**JULY-1998**	
93-0599	**437th AW**		
93-0601	**97 AMW**	**SEPT-1998**	
93-0604	**97th AMW**		
94-0065	**437th AW**	**MARCH-1999**	
94-0066	**437th AW**	**AUG-1998**	
94-0066	**437 AW**	**JAN-2000**	
94-0068	**437th AW**		
95-0102	**437th AW**		
95-0105	**97 AMW**	**DEC-1999**	
95-0106	**437 AW**	**JAN-2000**	***Spirit of Bob Hope***
95-0107	**437 AW**	**JAN-2000**	
96-0002	**437 AW**	**JAN-2000**	***Spirit of the Air Force***
96-0003	**437 AW**	**JAN-2000**	
96-0005	**437 AW**	**JAN-2000**	
96-0007	**437 AW**	**JAN-2000**	
96-0008	**437/515 AW**	**JAN-2000**	***Spirit of America's Veterans***
97-0044	**437 AW**	**JAN-2000**	
98-0049	**437/515 AW**	**JAN-2000**	
98-0050	**437 AW**	**JAN-2000**	